Studies in Computational Intelligence

Volume 618

Series editor

Janusz Kacprzyk, Polish Academy of Sciences, Warsaw, Poland
e-mail: kacprzyk@ibspan.waw.pl

About this Series

The series "Studies in Computational Intelligence" (SCI) publishes new developments and advances in the various areas of computational intelligence – quickly and with a high quality. The intent is to cover the theory, applications, and design methods of computational intelligence, as embedded in the fields of engineering, computer science, physics and life sciences, as well as the methodologies behind them. The series contains monographs, lecture notes and edited volumes in computational intelligence spanning the areas of neural networks, connectionist systems, genetic algorithms, evolutionary computation, artificial intelligence, cellular automata, self-organizing systems, soft computing, fuzzy systems, and hybrid intelligent systems. Of particular value to both the contributors and the readership are the short publication timeframe and the world-wide distribution, which enable both wide and rapid dissemination of research output.

More information about this series at http://www.springer.com/series/7092

Krzysztof Krawiec

Behavioral Program Synthesis with Genetic Programming

 Springer

Krzysztof Krawiec
Poznan University of Technology
Institute of Computing Science
Poznan
Poland

ISSN 1860-949X   ISSN 1860-9503   (electronic)
Studies in Computational Intelligence
ISBN 978-3-319-80171-1   ISBN 978-3-319-27565-9   (eBook)
DOI 10.1007/978-3-319-27565-9

Printed on acid-free paper

This Springer imprint is published by SpringerNature
The registered company is Springer International Publishing AG Switzerland

To my dearest wife Alex and our amazing daughters:
Faustyna, Dominika, and Michasia

Foreword

It is always encouraging to see research that I have followed each year for multiple years be elegantly motivated, synthesized and communicated in the form of a monograph. This book is no exception. Prof. Chris Krawiec herein ties together an impressive set of threads that were initiated at different times and later became more sophisticated, nuanced and woven together into a thesis about program synthesis and how it can be impressively advanced by leveraging newly revealable information about program behavior. I strongly recommend the book to those with introductory or advanced experience in genetic programming and other means of program synthesis. It opens a fresh door to graduate students in evolutionary computation seeking insights into the compelling but exasperating research topic that is genetic programming.

I was introduced to the idea of evolutionary-inspired automatic program synthesis in 1991 when, as a graduate student, I read an early paper by John R. Koza on "genetic programming". Excited, I devoted all the passion and energy of a wet-eared graduate student into trying to contextualize it with respect to adaptive search in general and with respect to representation in evolutionary algorithms. Over the years genetic programming advances in topics such as theoretical foundations, representation choices, bloat, symbolic regression and machine learning have kept my enthusiasm strong while also propelling the development of a vibrant, capable research community.

In 2013 I found myself lucky to host Chris at MIT as a short-term member of the ALFA group for some months of his sabbatical. I was eager to collaborate with him because he was making progress in a key "nut" of genetic programming that had been, to that point, hard to crack: how to use the meaning and semantics of a candidate program to influence program synthesis. The research community was routinely using impoverished metrics related solely to performance (rather than behavior) as a largely unquestioned design choice. Because of this genetic programming was not close to being a robust program synthesis approach, capable of reliably producing programs for problems of arbitrary difficulty.

Chris has aptly, in this monograph, pinpointed this design deficiency as "the evaluation bottleneck". In his GECCO 2013 publication with Jerry Swan, he has shown how one could use an "execution record" to improve the evaluation information of a program by detecting execution patterns using machine learning. I will always consider it good fortune that Chris and I were able to connect so well intellectually and work together during his sabbatical to develop an extension of that technique that won us the best paper award at GECCO'14. In this volume, Chris takes the behavioral perspective further, providing it with solid formal footing and combining with the novel concept of search driver.

A fundamental idea of the monograph is that programs are complex entities that behave in rich ways and their behaviors can be analyzed to make program synthesis more effective. Characterizing programs with scalar fitness that reflects only the final program outcome misses that opportunity. Potentially, all information resulting from program execution can be put to use to make program synthesis more scalable. With that, the horizon for genetic programming shifts dramatically. While it is imperative to remain modest about progress to date, the advances in this monologue allow us to scratch beyond the surface toward program synthesis 'in the large', i.e., synthesis of Turing-complete programs and solving difficult problems.

I am sure that readers will learn a lot from this volume as did I, enjoy!

MIT, Cambridge, MA,   *Una-May O'Reilly*
March 2015

Preface

The number of programmable devices in the world is currently in the order of billions and grows at an immense pace.[1] They operate on various hardware platforms, different operating systems, and feature an untold number of software modules written in hundreds of programming languages.

This ubiquity and versatility of programmable devices creates growing demand for software that needs to be commissioned, designed, written, tested, integrated, deployed, and maintained. No wonder software engineering nowadays belongs to the most demanded professional skills, also in the domains historically not associated with computer science (CS), like biology, medicine, or psychology. CS is an integral part of curriculum not only in universities, but also in high and elementary schools.

Despite the manyfold increase of the number of CS graduates in recent decades, the growing supply of CS professionals still does not meet the demand, even though software development is much more efficient nowadays and routinely supported by tools that the IT specialists in the past could have only dreamed of. Integrated development environments, debuggers, profilers, testing frameworks, and other computer-aided software engineering tools are indispensable in today's software engineer's toolbox. Yet, despite all these aids, software development is still a challenging and resource-demanding process, in part because contemporary software artifacts can be orders of magnitude more complex than they used to be in the past.

---

[1] There were 1.2 billion computers worldwide in 2011 and 6.7 billion mobile cellular subscriptions in 2013 (wolframalpha.com), and the number of PCs in use is likely to pass the 2 billion mark in 2015 (worldometers.info).

This book originates in the belief that the extent of computer support in software development can be pushed even further by means of the arguably the most advanced aspect of computer-supported software development, namely in automated *program synthesis*. Program synthesis offers the possibility of programs being more or less automatically generated from specifications given in various forms. Once of interest only to academics, it has recently gained momentum and we witness the dawn of its usage on a commercial scale (e.g., [45]). Systems capable of automated program synthesis are still in their infancy, but their capabilities are growing fast. Ultimately, program synthesis 'in the large' can help close the gap between the efficiency of human developers and the growing market demand, not to mention the possibility of solving a range of conceptually challenging and interesting problems.

For anyone familiar with computer programming, automation of program synthesis may seem unrealistic. No wonder many have expressed reservations about it. Edsger Dijkstra, the programmers' greatest guru, stated once that

> (...) computing science is – and will always be – concerned with the interplay between mechanized and human symbol manipulation, usually referred to as 'computing' and 'programming' respectively. An immediate benefit of this insight is that it reveals 'automatic programming' as a contradiction in terms. [27]

Though Dijkstra may appear to question program synthesis, a deeper insight suggests otherwise. Indeed, the above-mentioned interplay with a human is important, because in the beginning there must be an 'intent', i.e., somebody has to specify somehow what the resulting program should do. But the history of computer science demonstrates that intent can be expressed in many ways. In the early days of programming, humans were forced to express intent directly in low-level computer-readable machine languages. The subsequent generations of programming paradigms, through imperative, object-oriented, declarative and functional programming, pushed up that human-computer interface to higher abstraction levels. As a result, an implementation of a complex software artifact in modern programming languages may require a handful of statements. And the exploration of novel ways of intent specification continues. To process data in a spreadsheet, users can nowadays express their intent by giving examples of desired outcome [45]. Control flows can be specified visually, e.g., in the Scratch environment [119]. And it is only human imagination that could put an end to exploration of ways in which humans might 'commission' software.

In this context, program synthesis appears to be simply yet another stage in the above succession; an abstraction level where expressing intent does not resemble programming anymore. This is clearly the case when, for instance, the desired program behavior is given by a set of examples, an approach that is common in practice and followed also in this book.

We can be deliberately provocative and take these considerations even further, asking: does programming really need human intent in the first place? In the end, the universe is abundant with *emergent phenomena* that question the need of intent. Life on Earth is a clear manifestation of that: no living creature has been designed, yet life with its evolutionary adaptations meets, to a greater or lesser extent, the 'requirements' imposed by an environment.

The evolutionary methodology of genetic programming, the program synthesis methodology used in this book, demonstrates that programming can do away without humans. Programs can *emerge* as an outcome of interactions of a synthesis method with some form of 'environment'. The final result of such a process may be not different from a program designed by a human (and sometimes can outperform it in certain respects). The time has come to seriously consider the possibility of computer systems that autonomously program themselves. The growing body of literature on and achievements of contemporary program synthesis cited in this volume forms a strong evidence for this claim.

Nevertheless, this optimism does not obliterate the fact that program synthesis is a nontrivial task. For several reasons that will be detailed in this volume, the current capabilities of program synthesis methods are still quite limited. These limitations and other challenges form the main motivation for this book, which proposes several directions to address them.

Scope

The scope of this book is heuristic program synthesis, where the goal is to automatically generate a program that meets a given set of requirements, provided either as a set of *tests* (*examples*) or constraints concerning program input and output (a.k.a. *contracts*). We focus on the generate-and-test approach to program synthesis, of which the primary representative is nowadays genetic programming (GP) [79], a bio-inspired methodology of program induction based on the metaheuristic of evolutionary computation (EC).

Program execution involves complex interactions between the components of a program (e.g., instructions) and the data they operate on. Usually, only the final outcome of that process, i.e., program output, is used to guide the process of program synthesis. In GP, obtaining such a guidance usually involves aggregating the errors committed by a program on particular tests into a single scalar value that forms program evaluation. Such a design brings about 'evaluation bottleneck': the rich characteristics of the complex process of program execution are forced into a scalar value of evaluation function, which necessarily involves information loss.

We claim that relying exclusively on conventional evaluation function is more a habit than a necessity. The main motivation for this book is the observation that more information can be easily gathered from program execution and used to aid program synthesis. Evaluation has the potential of providing extensive information on *program behavior*; in the end, it is that behavior that determines if a program meets the requirements posed by a program synthesis task. Nothing precludes for instance scrutinizing program output for particular tests, or examining program actions on the level of individual instructions. The approaches presented here aim at better exploitation of such information for the sake of making program synthesis more effective. We achieve this is by characterizing programs with respect to multiple aspects and turning the single-objective problem of program synthesis (as it is normally posed in GP) into a multiobjective one.

To frame these ideas in a principled way, we come up with the concept of *search driver*, a formal object with at least minimal capability of guiding a search process. A search driver is not obliged to conform to the entirety of requirements normally imposed on evaluation functions, and, among others, may examine only selected aspect(s) of programs' behavioral characteristics. The main conclusion is that guiding search with several search drivers that are 'weak' in the above sense can be more efficient than using one conventional evaluation function. This thesis is corroborated with theoretical analyses and experimental investigations.

Contributions

The main contributions of this volume include:

- Identification of evaluation bottleneck in GP and analysis of its origins and consequences,
- The framework of behavioral program synthesis, unifying the developments in several threads of past GP research,
- The concept of execution record, a complete and universal account of program behavior for a given set of tests,
- The concept of search driver, a generalization of evaluation function and a source of 'weak' guideline for a search process, meant mainly to be used in connection with other search drivers,
- Methods of behavioral program synthesis that elicit more information from evaluated programs, broaden so the above-mentioned evaluation

bottleneck, and use the information acquired in this way to perform a better-informed and directed search,
- New experimental evidence for the viability of behavioral program synthesis.

A material upshot of this book is a publicly available software library that implements selected components presented in this book and is intended to ease further development of behavioral GP methods in the community (Chap. 10).

Organization and characteristics

This book provides a consolidated birds-eye perspective on the approaches to program synthesis that reach beyond the conventional GP template. To accomplish this mission and quickly engage the reader in discussion on the key issues, we spend relatively little time on introductions, keeping them at minimum. We start with providing the essential background for program synthesis in Chap. 1 and identifying the main challenges that limit the capabilities of contemporary program synthesis in Chap. 2. Chapter 3 introduces the formalisms for capturing program behavior and formulates the main postulate of this book in Sect. 2.6, providing so a footing and unified framework for the methods discussed in the following chapters.

In Chaps. 4–8, we take the reader on a journey though several behavior-based approaches to behavioral program synthesis, including implicit fitness sharing and related techniques (Chap. 4), semantic genetic programming (Chap. 5), trace consistency analysis (Chap. 6), and pattern-guided semantic programming (Chaps. 7 and 8). Most of these chapters report new or recent concepts and results; however, for completeness, we include also more mature approaches and show how they subscribe to are compatible with the vision of behavioral program synthesis.

Chapter 9 bridges the methods presented in Chaps. 4–8 by identifying and formalizing the concept of search driver. In doing so, it provides concrete tools to realize the behavioral program synthesis proposed in Chap. 3. In Chap. 10 we conduct an experimental comparison of the techniques discussed in the previous chapters (albeit smaller experiments are scattered across the book as well). We close this volume with 'grand vision' of behavioral perspective and discussion about its implications for program synthesis and beyond in Chap. 11, and final remarks and identification of promising avenues for future research in Chap. 12.

We find it appropriate to warn the reader what this book is *not* about. This monograph is not a textbook of program synthesis, nor a systematic review of the state-of-the-art of this discipline. Rather than that, it covers selected recent development in program synthesis, particularly as seen from the perspective of genetic programming. Neither should this volume be considered as a representative and unbiased source of knowledge of genetic programming or evolutionary computation. A reader interested in a broader perspective is referred to, among others, [106], [148], [117], and other comprehensive sources of knowledge in these disciplines. For a coverage of more recent advances in EC, we recommend [13].

As any other work on program synthesis, this book can be said to deal with *metaprogramming*: the algorithms considered here, once implemented, become computer programs that generate computer programs. We took special care to avoid confusion between these two levels of abstraction, consistently referring to the entities at the former level as 'algorithms', 'methods', while reserving the terms 'computer program' and 'program' exclusively for the latter level. Also, with GP being a special case of evolutionary algorithm, and an evolutionary algorithm being a metaheuristics, we find it appropriate to employ terminology common in all these areas. In that vein for instance 'programs' are 'candidate solutions' and the overall error committed by a program is an example of an 'evaluation function'. The reader is let known about these synonyms each time a new concept is introduced. However, we strive to avoid excessive reliance on bioinspired terminology and evolutionary metaphor, with the exception of a few well-grounded terms like 'mutation' and 'crossover'.

While taking care of precision of formalisms and notation, in most cases we do not explicitly delineate definitions from the continuous text. To ease the navigation, the key concepts are marked with margin notes and listed in the Index of terms.

Acknowledgements

This book is about soulless (and actually bodiless as well) machines, but everything that is told there originated in the minds of concrete people. This work would never see the light of day if it was not for my dear colleagues and collaborators, many of whom I am privileged to call my friends. By this token, my gratitude goes, in no particular order, to:

- Una-May O'Reilly of Massachusetts Institute of Technology, who accompanied me at the development of the key concepts of behavioral program synthesis,
- Jerry Swan of University of York, with whom we designed the conceptual framework of pattern-guided genetic programming,

- Alberto Moraglio of University of Exeter, with whom we developed the key concepts of semantic genetic programming and geometric semantic genetic programming,
- My postdocs and PhD students at Poznan University of Technology: Wojtek Jaśkowski, Bartek Wieloch, Marcin Szubert, Tomasz Pawlak, Karolina Stanislawska, and Paweł Liskowski, for sharing many great scientific adventures with me,
- Armando Solar-Lezama of Massachusetts Institute of Technology, with whom we 'transplanted' the pattern-guided program synthesis into the information theory context,
- Paweł Lichocki of Google, who helped me to develop the concept of cosolvability,
- Roman Słowiński of Poznan University of Technology, for continuous support and mentoring,
- Wolfgang Banzhaf of Memorial University of Newfoundland, Canada, for a great deal of inspiration and encouragement.

Credits go also to William B. Langdon of Univesity College London and Steven M. Gustafson of General Electric Research, who maintain the online genetic programming bibliography [105]. This invaluable repository was of immense help when researching for related work.

The research that has cumulated in the composition of this book has been financed from multiple funding sources, including the Polish-American Fulbright Commission, DS91507, EP/J017515/1, and Polish National Science Centre grant no. 2014/15/B/ST6/05205. This book has been written and typeset with LaTeX, and I implemented the software needed for experimental part of this work in Scala, an elegant and powerful programming language designed by Martin Odersky of École Polytechnique Fédérale de Lausanne and his team. I am indebted to the authors and contributors of these frameworks.

Poznań, Poland
August 2015

Krzysztof Krawiec

Contents

List of Acronyms

AST Abstract Syntax Tree

CoEA Coevolutionary Algorithm

DSL Domain-Specific Language

EA Evolutionary Algorithm

EC Evolutionary Computation

GP Genetic Programming

GSGP Geometric Semantic Genetic Programming

GSGX Geometric Semantic Crossover

GSGM Geometric Semantic Mutation

IFS Implicit Fitness Sharing

ILP Inductive Logic Programming

MAD Mean Absolute Deviation

MDL Minimum Description Length principle

ML Machine Learning

MSE Mean Square Error

NFL No Free Lunch theorem

NS Novelty Search

RL Reinforcement Learning

SGP Semantic Genetic Programming

TS Tournament Selection

TSP Travelling Salesperson Problem

1

Program synthesis

In this introductory chapter, we characterize and formalize the key concepts of this book, in particular computer programs. We also define the task of program synthesis and determine the main factors that make it challenging. Finally, we delineate several paradigms of program synthesis, among them genetic programming.

1.1 The nature of computer programs

Computer programs are unique among other mathematical formalisms in embodying *algorithms*, i.e. formal recipes for solving entire *classes* of problems. For instance, the greatest common denominator of *any* pair of integers can be calculated using the same short program. This makes programs fundamentally different from entities that are 'attached' to a specific problem *instance*, e.g., a specific route is a solution to a particular traveling salesperson problem.

Programs exhibit this characteristic because they are able to interact with *data*, or, in other words, respond to *input* with some *output*. This is actually more a necessity than an ability: programs *need* data to act upon. A program that expects an input cannot be launched without it. A deterministic program that does not take any input always produces the same output, which, apart from exotic usage scenarios[1], renders it useless.

A nontrivial program exhibits thus a spectrum of possible *behaviors* that depend on the input to which it is applied. Informally, it *does* something – a phrase that is hardly applicable to salesperson's routes. No wonder we tend to attribute programs with agency, saying that a program 'accepts',

---

[1] For instance, rather than storing a large raster image of a complex fractal, it may be more memory-efficient to store the program that generates that fractal – a compelling example of Kolmogorov complexity.

© Springer International Publishing Switzerland 2016
K. Krawiec, *Behavioral Program Synthesis with Genetic Programming*,
Studies in Computational Intelligence 618,
DOI: 10.1007/978-3-319-27565-9_1

'chooses', 'waits', 'assumes', 'guarantees', etc. Such anthropomorphisms feel natural and will by this token occur in this book, even though this habit has been sometimes criticized [27].

The expressive power of a program is conditional upon the programming language in which it is written. Any Turing-complete programming language is sufficient to express all computable functions, a class capacious enough to embrace most known problems of practical and theoretical interest. Even rudimentary programming languages are usually Turing-complete, including esoteric languages that, for instance, comprise just one instruction [39]. Programs written in such languages can implement most conceivable processes, from elementary arithmetic to simulating selected aspects of human intelligence. In particular, nothing precludes one from writing a program that manipulates other programs – interpreters, compilers, and virtual machines are natural examples of this capability.

program
instruction

In this book, by a computer program (*program* for short) we mean a finite discrete structure composed of elementary *instructions* (or *statements*) and capable of performing computation. The representation of programs that is most natural for humans is source code, i.e. text. For program synthesis, the textual form is redundant and cumbersome to handle, so virtually all approaches work with programs represented as *abstract syntax trees* (AST),

abstract
syntax
tree

abridged structures that contain only the effective elements of programs and omit, among others, the delimiters that separate syntactic structures in source code (like semicolons, parentheses, etc.).

program-
ming
language

The rules of forming *syntactically valid* (i.e. executable) programs from instructions in a given *programming language* are usually expressed as formal grammars. A grammar distinguishes the programs that belong to a given programming language from those that do not. In this book we consider only syntactically valid programs, and it will be sufficient for us to identify a programming language with a (possibly infinite) set $\mathcal{P}$ of programs and abstract from the particular formalism that determines their validity.

We write $p(in) = out$ to express that a program $p \in \mathcal{P}$ applied to an input data (*input* for short) *in* produces an output data *out* (*output*) as the result of execution. Inputs and outputs may be any formal objects representing certain *types*, either simple (usually scalars, e.g., bits, Booleans, numbers) or compound (usually data structures, e.g., lists, matrices, images). If program input is a tuple, its elements will be referred to as *input variables* and denoted by x_is, i.e. $in = (x_1, \ldots, x_k)$.

domain

The types associated with inputs and outputs determine the *domain* $(\mathcal{I}, \mathcal{O})$ of a program, where $\mathcal{I}$ and $\mathcal{O}$ respectively denote the sets of valid input and output values. The elements of $\mathcal{I}$ form *admissible inputsadmissible input*. For instance, the Boolean domain used in many examples throughout this book includes all programs with signatures of the form $\mathbb{B}^n \to \mathbb{B}$, i.e. $\mathcal{I} = \mathbb{B}^n$

and $\mathcal{O} = \mathbb{B}$, where $\mathbb{B} = \{\text{true}, \text{false}\}$. An input that does not belong to $\mathbb{B}^n$, e.g., a real number, is not admissible for programs in this domain.

As shown by Alan Turing [182], there is no way to determine in general whether a program terminates: the halting property is undecidable. For a non-halting program, it becomes impossible to verify if it returns the desired output for a given input, which is the key property in generative program synthesis (Sect. 1.3). To mitigate this problem, in this book we limit our interest to programs that halt. We also consider only deterministic programs.

1.2 Program synthesis

Writing computer programs is an activity that we habitually attribute to humans. In spite of this, the attempts to automate the process of generating computer programs, viz. *synthesize* them, date back to the early years of computer science and artificial intelligence (see, e.g., [188] and Sect. 1.5).

We define the task of *program synthesis* (*task* for short) as an ordered pair $(\mathcal{P}, Correct)$, where $\mathcal{P}$ is a programming language and $Correct : \mathcal{P} \to \mathbb{B}$ is a *correctness predicate*. Solving a task $(\mathcal{P}, Correct)$ consists in finding a program $p \in \mathcal{P}$ that fulfills $Correct$, i.e.: program synthesis task
correctness predicate

$$p \in \mathcal{P} : Correct(p), \tag{1.1}$$

(cf. [120]). A program p such that $Correct(p)$ is *correct* and forms a *solution* to a program synthesis task.

Because we adopt a metaheuristic approach to program synthesis (Sect. 1.5.3), it is important to explain how the notions introduced above relate to it. While program synthesis corresponds to *problem* (like the traveling salesperson problem mentioned earlier), a program synthesis task with a specific $\mathcal{P}$ and $Correct(p)$ corresponds to *problem instance* in metaheuristic terminology. The working programs considered by a running synthesis method are potential solutions and by this token are referred to as *candidate solutions*, *candidate programs*, or *search points*. A solution to a synthesis task corresponds to *optimal solution*, which we occasionally refer to as *optimal program*. candidate solution
optimal solution

The correctness predicate $Correct$ is responsible for telling apart the correct and incorrect programs in $\mathcal{P}$. As we detail later, there are several ways in which program correctness can be verified, i.e. *classes* of correctness predicates. For instance, the class exercised in this book involves confronting a program with tests. A given class of correctness predicate is instantiated by a *task specification* S, e.g., a specific set of tests. In most cases, the reference of $Correct$ to specification will be clear from the context and thus assumed implicit unless otherwise stated. Because we consider only halting programs, the correctness considered here is formally the *total correctness*. task specification

In algorithmic realization, the mathematical 'find such that' statement in (1.1) boils down to 'generate' or 'synthesize'. We find the latter term more adequate, as it emphasizes the fact that programs are assembled from smaller entities (instructions) and that programming is by nature combinatorial. This means that program synthesis lacks the concept of a *variable* and sets it apart from conventional optimization, where candidate solutions are usually fixed-length tuples of such variables. These arguments and past literature [44] incline us to lean toward the term *synthesis*.

discrete search problem

Posed in this way, program synthesis is a *discrete search problem* in the artificial intelligence (AI) sense [156], with *search states* are programs in $\mathcal{P}$. *Correct* partitions the search space into the *goal states* and the nongoal states, i.e. the programs sought and the remaining ones. In the most conservative formulation, this is the only source of information available to a method that performs program synthesis.

There is however an important feature that makes program synthesis a very special search problem. In conventional search problems, a goal test verifies an inherent property of a search state. For instance, to verify if a board state in the peg solitaire puzzle is a goal state, one checks if the number of pegs remaining on the board is one. In contrast, unless one reaches for *formal verification* methods (which are beyond the scope of this book), the correctness of a program cannot be determined by inspecting its structure, i.e. its source code. A program is correct if it *behaves* in the right way, i.e. if the $\mathcal{I} \rightarrow \mathcal{O}$ mapping it meets the requirements defined by *Correct*. Correctness of a program is intermediated by its interpretation (semantics), which is an *extrinsic* property, i.e. it is not explicitly present in the symbols that represent the instructions nor in their combination within a program. This behavioral aspect of program synthesis makes it nontrivial and will reverberate many times in this book.

solvable synthesis task

We consider only program synthesis tasks that are *solvable*. The necessary condition for a task to be solvable is that the programming language is expressive enough, i.e. a finite program that meets *Correct* can be formulated in that language, i.e. $\exists p \in \mathcal{P} : Correct(p)$. Expressibility of a programming language is however not sufficient to guarantee solving a given synthesis task *with a given method*. A synthesis algorithm can be inherently incapable of visiting some regions of search space due to, e.g., certain *search biases*.

1.3 Specifying program correctness

Program synthesis can be alternatively seen as *translation* of a specification S into a program $p \in \mathcal{P}$ such that $Correct_S(p)$. The key difference between these two entities is that specification is passive, i.e. can only be queried to determine program's correctness, while the resulting program is active in being executable.

A specification S defines the desired *effect* of computation, and as such can be conveniently expressed using *preconditions*, i.e. conditions that constrain the set of program inputs, and *postconditions*, i.e. conditions that program output has to meet given the input data. Formally, for a program $p : \mathcal{I} \to \mathcal{O}$ and a specification $S = (precond, postcond)$:

precondition

postcondition

$$p(in) \equiv out : postcond(in, out), \text{where } precond(in) \qquad (1.2)$$

where $precond : \mathcal{I} \to \mathbb{B}$, and $postcond : \mathcal{O} \to \mathbb{B}$. For instance, the specification of a program that calculates the integer approximation of the square root of a nonnegative number n can be phrased using pre- and postconditions as follows:

- $precond(n) = integer(n) \wedge n \geq 0$,
- $postcond(n, m) = integer(m) \wedge n^2 \leq m \leq (n+1)^2$.

Specifying program correctness by pre- and postconditions is common not only in theory [120] but also in practice, as epitomized by the growing popularity of the *design-by-contract* paradigm in software engineering [127, 133], where it is ofter realized using 'requires' and 'ensures' clauses.

design-by-contract

There are two fundamentally different ways in which pre- and postconditions can be verified for a given program. The *formal methods* achieve that without running the program, most commonly by constructing a formal proof of program correctness. The theorem to be proven is in general of the form:

$$\forall in : precond(in) \implies \exists out : postcond(in, out) \qquad (1.3)$$

If a *constructive* proof of such theorem can be conducted, it will also determine what is the *out* value that satisfies the postcondition. A side effect of conducting that proof is thus a synthesized program. This approach to program synthesis task is rooted in Hoare logic and formal verification [52] (cf. [120]).

Alternatively, the task can be specified by examples. In that case, the task specification S takes the form of a finite list[2] T of *tests*. Each test is an ordered pair (in, out), $in \in \mathcal{I}$, $out \in \mathcal{O}$, where in is program input, and out is the corresponding *desired output*.

test

desired output

We assume that T is *non-redundant*, i.e. $\nexists (in_1, out_2), (in_2, out_2) \in T : in_1 = in_2 \wedge out_1 = out_2$, and *coherent*, i.e. $\nexists (in, out_1), (in, out_2) : out_1 \neq out_2$. In genetic programming, tests are often referred to as *fitness cases*. A program synthesis task posed in this way can be considered as a machine learning (ML) task defined within the paradigm of *learning from examples*, with T

learning from examples

---

[2] In GP and machine learning literature, T is typically defined as a *set*. However, maintaining a fixed ordering of tests in T becomes important at certain point of our discourse, so we define T as a list.

playing the role of a *training set* and each test corresponding to an example. As in ML, T is not necessarily assumed to enumerate all possible program inputs; in general, it may be considered a sample drawn from a (potentially infinite) *universe* of tests $\mathcal{T}$.

Given a set of tests $T \subseteq \mathcal{T}$, we can define the correctness predicate as

$$Correct_T(p) \iff \forall(in, out) \in T : p(in) = out, \qquad (1.4)$$

where $p(in)$ denotes an application of program p to an input data *in*. The vector of desired outputs *out* is alternatively referred to as *target*.

target

Specifying correctness with examples is usually *partial*, because the desired behavior is unspecified for any $in : \nexists(in, out) \in T$. Formal specification is, by contrast, usually *complete* and thus more general. Yet, formal correctness predicates can be difficult to design without a strong mathematical background, and can sometimes be more difficult than writing the program in question. On the other hand, though reasoning in terms of examples is natural for humans, a large number of examples may be required to specify the desired behavior. In the search for alternatives, the notion of specification is being recently extended to embrace other ways of expressing the desired outcome of a synthesis process. In this context, program synthesis can be rephrased more generally as the task of discovering an executable program from *user intent* [44]. Recent interesting developments in this area include expressing intent interactively [45] and writing incomplete programs to be complemented by a synthesis system [167].

We propose to group program synthesis paradigms with respect to the workflow they implement. In the top-down *specificiation-driven* approach, it is the specification that drives the synthesis process. A synthesis algorithm starts with the given specification S, analyzes it, and derives (usually deduces) a program from it. The derived program conforms by construction to the *Correct* predicate, so it does not have to be verified for correctness. Such a workflow is characteristic to, among others, systematic deductive approaches to program synthesis [120] (see Sect. 1.5).

specifi-
cation-
driven
program
synthesis

In the bottom-up, *generative*, or *generate-and-test* approach, the synthesis process uses a generator of programs that works in a more 'undirected' way. The generated candidate programs are verified using the *Correct* predicate (which in such case can be considered as a form of *oracle*). The feedback from the verification is subsequently used to produce the next, hopefully better, candidate solution(s). Such generate-and-test workflow is of, among others, genetic programming [79, 148], where an evolutionary algorithm serves as a generator of programs, and program correctness is verified by an evaluation function (Sect. 1.5.3).

generative
program
synthesis

An implication of adopting the top-down mode is that a program synthesis method has to 'understand' the specification in order to translate it into a program. In contrast, such a capability is not essential for bottom-up, generative approaches. The latter are thus more domain-independent, and can be conveniently used with complex domain-specific languages, where instructions may be intricate and have complex effect; the languages designed for image analysis may serve as examples here [11, 90]. In a sense, generative approaches assume that the synthesis task in question is too complex to be solved analytically and has to be heuristically 'datamined' to gain some understanding of it, and so facilitate finding a solution. This perspective is congruent with our vision of behavioral program synthesis (Chap. 3), and is one of the reasons why this perspective is built upon the generative stochastic metaheuristic of genetic programming.

1.4 Challenges in program synthesis

There are several reasons why program synthesis is challenging and robust and scalable program synthesizers are yet to be seen. The most obvious one is the size of the search space. The number of combinations of instructions grows exponentially with program length, even if only some of them are syntactically correct in a given programming language. This affects not only the bottom-up methods that need to search that space directly, but also indirectly the top-down approaches, because the size of program space is reflected in the number of paths in the proof space that need to be considered.

As an example, consider the task of synthesizing an m-ary Boolean function $\mathbb{B}^m \rightarrow \mathbb{B}$ represented as an expression tree, in which the programming language comprises k binary (i.e. two-argument) instructions. There are $\binom{n/2}{k}\binom{n/2}{m}cat(n)$ programs represented as trees composed of n instructions, where $cat(n)$ is the nth Catalan number: $cat(n) = \binom{2n}{n}/(n+1)$ [180]. For simplicity, let us assume that the task can be solved using a program that fetches each of m input variables exactly once, i.e. such that forms a binary tree with m leaves and $m - 1$ internal nodes. Even for the moderately difficult 11-bit multiplexer ($m = 11$) [122] and $k = 4$ binary instructions, the above formula results in the staggering 2.93×10^{11} programs – and this is a *conservative estimate* of the size of the search space that needs to be explored.

The second challenge in program synthesis is that programming languages are rich enough to express the same functionality in many ways. Formally, the mapping from the space of programs $\mathcal{P}$ to the space of their behaviors (interpreted for instance as the outputs produced for all tests, like in semantic GP, Chap. 5) is many-to-one. This non-injective characteristic manifests also in the existence of multiple correct programs for a given task, or, put in terms of search problems, in the existence of multiple goal states. When a search problem of program synthesis is recast as an optimization multimodal evaluation function

problem (Sect. 1.5.3), this causes an evaluation function to be *multimodal*. However, this is contingent also on the structure of the search space induced by search operators, as accurately commented for evolutionary algorithms (EA) by Lee Altenberg:

> The multiple-attractor problem is usually described as "multimodality" of the fitness function, but it must be understood that the fitness function by itself does not determine whether the EA has multiple domains of attraction – it is only the relationship of the fitness function to the variation-producing operators that produces multiple-attractors. [2, p. 4]

On one hand, multimodality increases the statistical odds of finding a solution; on the other, it makes it more difficult to prioritize the search in the presence of multiple potentially useful search directions. Also, multimodality may be a sign of a program synthesis task being *underconstrained*, which is particularly likely when a correctness predicate involves few tests. In such cases, the

generalization

synthesized program is expected to *generalize* well beyond the training set of tests. This presents a challenge on its own: how to ensure, for instance, that a program meant to calculate the median of a list of numbers, synthesized from a handful of tests, calculates the correct value for any input?

Expressibility of computer programs gives rise to yet another problem. Instructions, the underlying components of programs, are abstract symbols that do not mean anything on their own. Their meaning resides in *semantics*,

semantics

materialized in the 'substrate' that provides for program execution (be it an interpreter, compiler, or hardware). The semantics of individual instructions is usually simple, and in generic programming languages may comprise little more than elementary logic and arithmetic. Yet, because instructions can be applied in different contexts (e.g., to various arguments, variables, subprograms, etc.) and in various orders, their overall effect is hard to model. As a consequence, the impact of a given instruction on the final computation outcome is highly contextual – the interpretation of a given piece of code in a program depends on its surroundings. This is particularly evident in imperative programming languages (and virtually absent in the functional ones). Put in terms of evolutionary computation, the underlying vehicle of GP, if instructions in a program are likened to genes in a chromosome, then there

epistasis

is strong *epistasis* between them (cf. Sect. 1.5.3).

The complexity and essential character of interactions between instructions is such a prominent feature of programs that it inclined John H. Holland

emergence

to use them to illustrate emergence in his seminal work on this topic:

> Interactions play a central role in the study of emergence. A detailed knowledge of the repertoire of an individual ant does not prepare us for the remarkable flexibility of the ant colony. The capabilities of a computer program are hardly revealed by a detailed analysis of the small set of instructions used to compose the program. [56, pp. 38-39]

Indeed, as we argued elsewhere [84, 5], a behavior of a program can be considered its emergent property. Only a part of behavior (pertaining to the program's final outcome) matters for solving a synthesis task; what a program does on its route to producing an outcome is – in a sense – irrelevant. Arbitrarily complex behaviors of programs emerge from a handful of relatively simple instructions. In this light, it is not an overstatement to equate a program, or even more a yet-imperfect program 'in the making', with a complex system [131].

It may be useful at this point to confront the complexity of semantic effects of program execution with a conventional AI-type search problem, the peg puzzle mentioned in Sect. 1.2. The effects of peg moves (modifications of the current solution) directly follow from the board structure and state. They do not refer to any external 'body of knowledge', like semantics of logic or arithmetic instructions in programming. Similar moves (e.g., moves applied to the same peg) have usually similar effects. The evaluation function (the number of pegs left on the board) changes more or less gradually with consecutive moves. Compared to computer programs, this is a really straightforward environment.

The above-mentioned property of similar moves having similar effects is closely related to the notion of *locality* in evolutionary computation (EC) [154]. A problem is said to exhibit high locality if applying a move to a candidate solution leads to solution with similar evaluation. High locality facilitates designing search operators and is usually considered as a sign of a problem's simplicity. Consider a conventional optimization task, the combinatorial traveling salesman problem (TSP). A candidate solution in TSP is an ordering of cities to be visited, encoded as a permutation of natural numbers. Similar permutations in TSP tend to represent similar routes. A move that swaps two edges in a route may affect route length but will not 'ruin' it, as all the remaining edges remain intact.

locality

Computer programs are notorious for being anything but local in the above sense [34, 85]. The mapping from program code to its behavior can be particularly complex: a minute modification of the former may cause a dramatic change in the latter. On the other hand, a major change in a program may be behaviorally neutral. In other words, conventional objective function in GP is known to exhibit low *fitness-distance correlation* [181], i.e. it does not correlate well with the measures of syntactic similarity between programs. This applies to generic distance measures like edit distance (see, e.g., [140]) as well as to operator-based distance measures, like crossover-based distance proposed in [46]. Put in yet another way, fitness landscapes in GP tend to be 'rugged'.

fitness-distance correlation

These properties of programs has been aptly commented by Edsger Dijkstra:

In the discrete world of computing, there is no meaningful metric in which "small" changes and "small" effects go hand in hand, and there never will be. [27]

James Gleick phrased this characteristics in a more general way, but also more evocatively:

Computer programs are the most intricate, delicately balanced and finely interwoven of all the products of human industry to date. They are machines with far more moving parts than any engine: the parts don't wear out, but they interact and rub up against one another in ways the programmers themselves cannot predict. [40, p. 19]

In summary, program synthesis is a challenging task due to size of a search space, its multimodality, externalized semantics of instructions, and complex interactions between them. It is thus not surprising that it spawned not one but several research paradigms presented in the next section.

1.5 Paradigms of program synthesis

In this section, we characterize the main paradigms of program synthesis: deductive program synthesis (Sect. 1.5.1), inductive programming (Sect. 1.5.2), and genetic programming (Sect. 1.5.3), the approach used in this book. The former two paradigms are largely top-down according to the taxonomy introduced in Sect. 1.3, while the latter one is purely bottom-up and generative. Rather than providing a complete review, our aim in this section is to position genetic programming in the context of other paradigms.

1.5.1 Deductive program synthesis

deductive program synthesis

In *deductive program synthesis*, one assumes that task specification is complete, i.e. determines the desired output of a sought program for all admissible inputs. The cornerstone of this paradigm is the Curry-Howard correspondence [59], which proves a one-to-one relationship between programs in computer science and proofs in logic. By this virtue, deductive program synthesis boils down to theorem proving, and involves transformation rules, unification, and resolution [120].

The key advantage of deductive synthesis is that the resulting programs are *correct by construction* [28]. On the other hand, its usefulness directly depends on effectiveness of theorem provers, which is nowadays still quite limited. Moreover, achieving complete proof automation is challenging; this is one of the reasons why, for instance, the Coq system, which famously helped proving the four-color theorem, is advertised as a 'proof assistant' rather than a 'theorem prover' [32]. As a consequence, deductive synthesis approaches do not scale well and, depending on the genre, are currently capable of synthesizing programs no longer than a few dozen instructions.

The other challenge for deductive program synthesis stems, paradoxically, from its complete nature. Specifying the desired behavior for all possible inputs is natural for more formal program synthesis tasks, like the square root function considered in Sect. 1.3. However, for many tasks the desired behavior may be not explicitly given. Consider for instance a program that implements a game strategy and should respond with an action (output) to a given board state (input). As the ultimate game outcome is delayed and conditional upon the behavior of an opponent, the most desirable move (desired output) may be not known for a give board state.

Last but not least, even if a complete specification of behavior does exist, it may be cumbersome or difficult for a human programmer to formalize it. It may be thus more natural to express the desired outcome of program synthesis by, e.g., providing examples of desired behavior. Such a process is characteristic of inductive programming as described in the next section.

1.5.2 Inductive programming

Contrary to deductive program synthesis, in *inductive programming* task specification is not assumed to be complete: admissible inputs to a program may exist for which the corresponding output is not given. Specification has the form of a list of tests T, which do not have to enumerate all admissible program inputs (1.4). A synthesis method is expected to perform *induction*, i.e. synthesize a program that does not only passes the tests in T, but also behaves 'accordingly' for the inputs not covered by task specification, i.e. for tests in $\mathcal{T} \setminus T$, where $\mathcal{T}$ is the universe of all tests (cf. Sect. 1.3). What 'accordingly' means depends on the given task and domain, and is often not formalized. For instance, given only a handful of examples of people's full names, a synthesized program may be expected to correctly extract the initials for *any* first, middle, and last name [45].

inductive program-ming

Such formulation of program synthesis entails *generalization* and clearly resonates with learning. Indeed, the primary representative of inductive programming is *inductive logic programming* (ILP, [162, 161]), recognized nowadays as a branch of machine learning (see, e.g., [134, Ch. 10]. ILP deals mostly with logic-based programming languages, in particular Prolog. Main research efforts in ILP focus on learning from relational data, knowledge discovery, and data mining.

inductive logic program-ming

program
synthesis
vs.
machine
learning
Inductive program synthesis bears also a certain similarity to *learning from examples*, the arguably most popular paradigm of machine learning [134]. In a sense, program synthesis subsumes machine learning, as every (realizable) classifier can be (and usually is) implemented as a computer program. In this context, a machine learning induction algorithm can be treated as a special form of program synthesizer. Nevertheless, the roads of program synthesis and the mainstream of ML parted ways in the 1990s. ML focused on specific (and often non-symbolic and thus non-transparent) representations of hypotheses (like decision trees, decision rules, bayesian networks, etc.), and in exchange for that enjoyed the availability of efficient (though usually heuristic) synthesis algorithms for inducing them. Program synthesis, on the other hand, could not sacrifice its generality (and transparency of the programs) without losing its primary mission. With the advent of strongly non-symbolic paradigms in ML (e.g., support vector machines and more recently deep neural networks), this chasm only got deeper, and today few consider program synthesis as a form of ML.

1.5.3 Genetic programming

genetic
program-
ming
Genetic programming (GP) is a stochastic generate-and-test approach to inductive program synthesis [79, 81]. It rephrases program synthesis as an optimization problem and relies on the metaheuristic of evolutionary algorithms [160, 33, 54], arguably one of the few key bio-inspired metaheuristic approaches [168], to iteratively improve candidate programs. Remarkably, GP has been an important paradigm of EC from the early days of this discipline and many pioneering EC studies were dedicated to evolution of *executable structures*[3]. For instance, as emphasized by Mitchell [131, chap. 9], much of John Holland's early work on rule systems [55] was driven by the urge to evolve executable objects.

GP shares its architectural underpinnings with other incarnations of the evolutionary metaheuristic, like genetic algorithms (GA) and evolutionary strategies (ES). This iterative search procedure, shown in Fig. 1.1, maintains a working set of candidate solutions P called *population*. The elements of P are programs (candidate programs) and are sometimes referred to as *individuals*. Initially, P is populated with randomly generated candidate programs from the programming language of consideration, i.e. $P \subset \mathcal{P}$. The quality of each program $p \in P$ is then assessed using an *evaluation*
fitness
function (which we will also occasionally call *fitness* for consistency with past work). If evaluation reveals an optimal program p^*, the search is terminated and p^* is returned as the outcome. Otherwise, a *selection operator*

---

[3] An executable structure needs to interact with some external 'stimulus' for its characteristics to be revealed. This definition embraces conventional programs (Sect. 1.1), but also for instance analog circuits studied by Koza [81].

is applied to P, producing a subset $P' \subseteq P$ of most promising programs called *parents*. Next, *search operators* are applied to the elements of P', resulting in *offspring* candidate programs, which form the next population P to be processed by the subsequent iteration of the evolutionary loop.[4]

What follows then from this description and from Fig. 1.1 is that an evaluation function plays the decisive role in a GP synthesis process. It is in the center of focus of this book and we will come back to it later in this section.

Apart from evaluation, the course of an evolutionary run is determined by a selection operator and search operators. A typical *selection operator* has the signature $sel : 2^{\mathcal{P}} \to \mathcal{P}$ and, when applied to a working population P, selects a well-performing individual from it. In this book we use only *ordinal* selection operators that interpret evaluation as a value on an ordinal scale (not necessarily a metric scale). Such operators can be alternatively termed *non-parametric* [117, p. 45]. The default selection operator in GP is *tournament selection* (TS). TS samples a low number k (usually $k \in [2, 7]$) of candidate solutions from the population and returns the best of them. It became the common method of choice in GP when it has been recognized that selecting solutions proportionally to fitness (*fitness-proportionate selection*) makes it likely for the best-performing candidate solution to dominate the entire population.

selection operator

ordinal selection operator

tournament selection

fitness-proportionate selection

Search operators are typically unary (*mutation*, $\mathcal{P} \to \mathcal{P}$) or binary (*crossover*, $\mathcal{P} \times \mathcal{P} \to \mathcal{P} \times \mathcal{P}$). The role of the former is to introduce minor changes in candidate solutions; the latter should *recombine* the parent solutions so that the offspring share certain 'traits' with them. For instance in so-called tree-based GP (detailed further in this section), mutation may replace a piece of the parent's AST tree with a randomly generated tree, and crossover swap two randomly selected subtrees in parents' ASTs. In GP, mutation and crossover are often used in parallel, so that some offspring stem from the former while some from the latter. In EC terms, these operators together are supposed to provide for *variation*, which, along with selection, forms the two cornerstones of evolution. All these operators are usually stochastic, i.e. two applications of an operator to the same population P will usually result in a different outcome.[5]

search operator

A GP algorithm thus performs a parallel, population-based search, and is by this virtue expected to be relatively resistant to the risk of gravitating to and getting stuck in local optima. For this reason, it subscribes to the category of *global search*.

The above 'vanilla GP' can be modified and extended in many dimensions, for instance by updating the population individual-by-individual (called

---

[4] Populations and other collections of candidate solutions are formally multisets, but we refer to them as 'sets' for brevity.

[5] Technically, they are thus random variables or, more precisely, *random functions*.

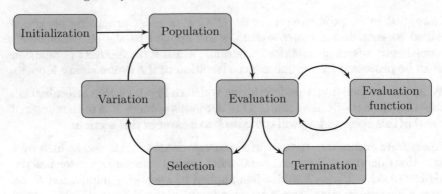

Fig. 1.1: Conventional workflow of genetic programming.

steady-state evolution in contrast to the above *generational evolution*), involving elitism, partitioning the population into *islands*, maintaining an internal *archive* of well-performing candidate solutions, not to mention the panoply of sophisticated selection and search operators. The reader interested in such extensions is referred to textbooks on GP [7, 81, 148] and the online bibliography of GP papers [105].

All those components, however important and beneficial for GP performance, are largely beyond the scope of this book, as our main focus is on the evaluation function, which is arguably the 'root cause' of most decisions made by a search algorithm. In GP, evaluation is based on the performance of a candidate program, i.e. its conformance with the desired behavior as specified by program synthesis task. However, the original formulation of program synthesis as a search problem (1.4) cannot be directly implanted into GP. Evolution, whether natural or simulated, is all about *accretion*, i.e. gradual accumulation of improvements that give individuals a reproductive advantage. It is thus typically assumed that an evolutionary algorithm needs a continuous, or at least multi-valued measure of a solution's quality, i.e. *fitness*, to drive the iterative improvement process. Therefore, virtually all GP genres abandon the qualitative correctness predicate *Correct* in favor of *evaluation function* f with a codomain defined on a scale that is at least ordinal, and usually real-valued, i.e. $f : \mathcal{P} \to \mathbb{R}$. Without loss of generality, we will assume that f is minimized (if not stated otherwise), even though this is somehow inconsistent with the etymology of the term 'fitness'. Nevertheless, to minimize abuse of biological metaphor [168] and for the more fundamental reasons we discuss in Sect. 2.5, we will restrain from using the term 'fitness' unless it is historically justified.

evaluation function

consistent evaluation function The evaluation functions used in GP are usually consistent with *Correct*, i.e. can indicate an arrival at an optimal solution:

$$f(p) = 0 \iff Correct(p). \tag{1.5}$$

In other words, under a consistent evaluation function, the notion of optimal solution converges with the notion of correct program. Given a consistent evaluation function f, solving a solvable program synthesis task with GP boils down to finding such p^* that

$$p^* = \arg\min_{p \in \mathcal{P}} f(p). \tag{1.6}$$

Contrary to popular belief, we claim that it is not obvious what is the 'right' evaluation function for a given task (or even class of tasks). The formulation of program synthesis (1.1) is agnostic about that. Given a program synthesis task, there will be usually infinitely many evaluation functions that are consistent with its correctness predicate. This observation is important for this book and will ultimately lead us to the concept of *search driver* presented in Chap. 9.

Nevertheless, it is commonly agreed that an evaluation function f should express a program's 'degree of correctness'. GP methods typically calculate such a degree based on program's behavior on tests. Most commonly, f takes the form of

$$f_o(p) = |\{(in, out) \in T : p(in) \neq out\}| \tag{1.7}$$

where T is a nonempty finite list of tests (Sect. 1.3). f_o counts thus the number of tests *failed* (not passed) by p. Alternatively, $f_o(p)$ may count the tests passed by p (a quantity known also as the *number of hits*), in which case it would have to be maximized. In either case, f_o is intended to capture the 'absolute' quality of a program, and by this token is in the following referred to as *objective function*. An objective function is the evaluation function that 'comes with the problem' and is in this sense recognized as the appropriate assessment method of candidate solutions. It is used in GP by default, and to emphasize this fact we will alternatively refer to it as *conventional evaluation function*.

objective function

conventional evaluation function

As signaled in Sect. 1.3, T is often taken from a larger (and sometimes infinite) universe of tests $\mathcal{T}$ and forms in this sense a *training set*. (1.7) becomes in such cases an estimate of the 'true' underlying evaluation, i.e. the fraction of tests passed in entire $\mathcal{T}$.

GP turns the original search problem of program synthesis into an optimization problem. The means by which this is achieved is the *relaxation* of the binary correctness predicate (1.4) into an ordinal evaluation function, in the canonical case f_o. In consequence, GP allows programs to be 'partially correct'. Behind this apparent oxymoron, there is evolutionary rationale related to the aforementioned accretion. Programs that pass only some tests can be iteratively improved and ultimately become correct. Also, exact conformance with the original specification is not critical in some domains. A canonical example is *symbolic regression* , where GP seeks a nonlinear regression model by synthesizing programs that operate in a (typically)

symbolic regression

real-valued domain (i.e. here $(\mathcal{I}, \mathcal{O}) = (\mathbb{R}^n, \mathbb{R})$). The evaluation function commonly used for solving symbolic regression problems with GP is the mean square error (MSE), equivalent up to ordering of candidate programs to the Euclidean distance[6]:

$$f_E(p) = \sum_{(in, out) \in T} (p(in) - out)^2. \tag{1.8}$$

Confronting this formula with f_o (1.7) reveals that f_E 'fuzzifies' the concept of passing a test. This observation will become relevant when defining test-based problems (Sect. 4.1) and program semantics (Chap. 5).

Because we assumed earlier that $\mathcal{P}$ hosts all candidate programs of interest, no additional constraints are necessary to delineate the search space in (1.6), which makes it is an *unconstrained optimization task*. If, for instance, task formulation requires the program being sought to not exceed certain length, we assume that all programs in $\mathcal{P}$ by definition meet such a constraint.

The way in which a GP algorithm navigates a search space of programs is in part determined by how they are represented. Past GP research delivered several alternative program representations. There is the conventional *tree-based GP*, where programs are represented as expression trees [79], usually equivalent to ASTs. There is the *linear GP*, where programs are sequences of instructions [6, 14]. Another program representations are nested lists of instructions that operate on stacks (*PushGP*, [170]) and graphs of instructions, with edges determining the dataflow between them (*Cartesian GP*, [129]). All these approaches vary only in program representation and conform to the formalisms introduced above.

tree-based GP

It should become clear at this point that GP is a methodology that reaches well beyond program synthesis. In contrast to typical formal methods, GP can for instance handle imperfect task formulations (e.g., inconsistent tests) or noisy data. As a consequence, the list of human-competitive achievements of GP is impressive [80, 73]. It is commonly believed that GP's capabilities stem from a combination of two key elements. The first is representing candidate solutions as programs, either conventional or algorithms for classification, regression, clustering, reasoning, problem solving, feature construction, etc. This flexibility enables expressing solutions to virtually any type of problems, whether the task in question is learning, optimization, problem solving, game playing, etc. The second key element is the reliance on the 'mechanics' borrowed from biological evolution, which is unquestionably a very powerful computing paradigm, given that it resulted in life on Earth and development of intelligent beings. This hypothesis, though never scrupulously verified to date, seems to be propelling the interest in and progress of GP.

---

[6] A GP run that employs tournament selection or other ordinal selection operator will proceed identically for MSE and the Euclidean distance.

1.6 Consequences of automated program synthesis

Once one realizes the capacities of computer programs, it does not take long to notice that the potential consequences of automated program synthesis 'in the large' are profound. Automatically synthesized programs would elevate the robustness of software and implicitly, that of many other technologies. Provably correct programs would make software certifiable, which nowadays can be realized on a very limited scale and only in certain contexts using, e.g., the Coq proof assistant [32]. Automatically generated software would be cheap to produce and malware-free. It could be also paramount with respect to non-functional properties like runtime, memory footprint, or power consumption.

Remarkably, these benefits would stretch beyond the boundaries of programming as currently practiced by humans. Automated program synthesis could help solving tasks that are nowadays either conceptually too complex to tackle, or economically not viable. A particularly useful application is synthesis in programming languages that are difficult and cumbersome for humans but used in practice for all sorts of reasons (legacy, efficient translation into machine language, etc.).

The future of program synthesis can be to some extend foretold by the telltales of current developments. In the following, we touch upon two areas of program synthesis that witnessed remarkable progress in recent years and seem particularly promising: program improvement and end-user programming. It goes without saying that this choice is subjective and other avenues exist, but their full coverage is beyond the scope of this book.

1.6.1 Program improvement

Because synthesizing programs from scratch is challenging (Sect. 1.4), we recently witness growing interest in methods that aim at *improvement* of programs written by humans, more specifically of their non-functional properties like runtime, memory occupancy or power consumption. The key advantage is that a human-written reference program determines the target of the synthesis process. It can be used as a test generator to construct a program synthesis task, or serve as a source of task specification, which can be derived from it using formal methods (e.g., [19]). The former usage is particularly valuable when supply of tests is limited, which is common in some branches of program synthesis [45].

Improvement of non-functional properties has been approached on various abstraction levels. On the level of machine language, it relates to *rewrite systems* studied in compiler design and code optimization. For instance, in [158], Schkufza et al. employed the Markov Chain Monte Carlo technique to improve the runtime of programs written in machine code for a 64-bit

<small>program improvement</small>

<small>non-functional properties</small>

<small>rewrite systems</small>

x86 processor. The reference program is machine code written by a human or compiled from a higher-level language. The Metropolis-Hastings algorithm is used to stochastically generate new candidate programs from it. The authors employ search operators similar to mutations in GP, randomly modifying instructions' opcodes, operands, replacing entire instructions, or swapping them within a program. The optimization is driven by an evaluation function that returns a weighted sum of estimated program runtime and Hamming distance between the desired and actual output for a set of tests. The experiment conducted on the benchmarks taken the famous *Hacker's Delight* volume [189] show an almost systematic reduction of runtime (up to 40 percent), often accompanied with shortening of the resulting code (e.g., from 31 to 14 lines in the case of of Montgomery multiplication procedure). Remarkably, the observed speedups improve over the conventional compilers run with the most intense optimization (e.g., gcc -O3).

At a higher abstraction level, Langdon et al. developed a GP framework for manipulating source code written in C++ and applied it to several domains. In [146, 147], they optimized the code of MiniSAT, a popular Boolean satisfiability (SAT) problem solver and obtained accelerations of execution greater than those elaborated by human programmers. In [107], they achieved up to six-time reduction of execution time of a computer vision procedure (stereo disparity estimation algorithm) written for the CUDA architecture running on GPUs. In [108], they reported over 35 percent speedup of registration procedures for magnetic resonance imaging.

At an even higher abstraction level, Kocsis and Swan [77] proposed a more formal method that operates on ASTs and exploits the knowledge of data types to improve programs. By making use of the well-known Curry-Howard isomorphism between proofs and programs [59], they replaced a (traditionally stochastic and non-semantics-preserving) GP mutation operator with deterministic proof-search in the *sequent calculus*. They showed how this operator can be used to automatically replace the singly-linked implementation of a list with the more efficient implementation of a difference list. On the implementation side, they used the reflection mechanism built-in to the Scala programming language to search for amenable data types and accordingly modify the AST trees of the original source code. The semantics-neutral character of this method makes it potentially applicable not only in GP (and in typed GP in particular), but also in the formal and deterministic methods of program synthesis.

1.6.2 Hybrid and interactive program synthesis

In its canonical formulation, program synthesis proceeds in an 'off-line' mode: a user prepares the specification, chooses the programming language (or designs an ad-hoc one), passes them to the synthesis method, and waits for a program to be synthesized. At the current state of advancement of program synthesis, such usage scenario turns out to be far from realistic for programs longer than toy examples. As contemporary techniques do not scale well, preparing a specification and program synthesis may together require more time than writing the program manually.

In response to this, *hybrid* and *interactive* approaches to program synthesis have recently gained more attention. An example of the former can be *sketching* [167], where a user writes a *partial program*, i.e. a program that is missing pieces of code while being otherwise syntactically correct. The method fills in the gaps with pieces of code that complement the partial program so that it becomes correct. By sharing the process of program between a human and a machine, sketching intends to lower the computational complexity of program synthesis.

Interactive approaches to program synthesis assume that a human operator is willing to aid the synthesis process at selected stages. This is particularly useful in *end user programming*, intended to support users with limited programming capabilities. In such application scenarios, one often cannot assume that a user has *any* level of understanding of programming languages. A recent example is here Flash Fill [45], a technology recently developed at Microsoft™ Research and deployed in the 2014 edition of Microsoft Excel™. Flash Fill allows a user to specify a desired transformation of data in a spreadsheet by providing a few examples of what is the desired effect of that transformation. Based on those examples, Flash Fill synthesizes an ad-hoc data transformation program in a domain-specific language, and applies that program to all data entries. By inspecting the outcome and possibly correcting it, the user provides a more detailed feedback for the method, which is used to fine-tune the synthesized program. Internally, Flash Fill relies on a carefully customized domain-specific programming language and uses machine learning techniques to select the hypotheses (candidate programs) that are most likely to meet user expectations.

end user programming

1.7 Summary

In this chapter, we characterized the key properties of programs, presented and formalized the task of program synthesis, and delineated its main paradigms. We also identified the main challenges one faces when attempting to synthesize programs automatically. These challenges limit the capabilities of program synthesis methods. However, we claim that this is in part due to certain design choices that are commonly followed in the generative methods like GP. In this book we focus on the limitations pertaining to the way a search algorithm is informed about the qualities of working solutions. The next chapter is entirely devoted to this aspect.

2

Limitations of conventional program evaluation

In Sect. 1.4, we identified several challenges for program synthesis, among others the vastness of search space and the intricate way in which program code determines the effects of computation. In this chapter, we identify and discuss the consequences of the conventional approach to program evaluation in generative program synthesis. Though focused mostly on the generative paradigm of GP, some of the observations made here apply to other generative synthesis approaches.

2.1 Evaluation bottleneck

As introduced in Sect. 1.5.3, the evolutionary search process in conventional GP is driven by an evaluation function (fitness) $f(p)$ applied to candidate programs $p \in P$. Evaluation typically boils down to counting the number of tests failed or passed by p. For convenience we repeat here the formula for the conventional evaluation function, i.e. the objective function (1.7):

$$f_o(p) = |\{(in, out) \in T : p(in) \neq out\}|. \qquad (2.1)$$

Assessing candidate solutions using a scalar performance measure has several merits. First of all, it is in a sense minimal – it is hard to imagine a simpler way of informing a search algorithm about the solutions' characteristics. It is also compatible with the conventional way of posing problems in optimization and machine learning. Last but not least, it eases separation of generic search algorithms from domain-specific evaluation functions, which is so essential for *meta*heuristics. No wonder that this 'design pattern' is so common that we rarely ponder over its other consequences.

Unfortunately, there is a price to pay for all these conveniences, a price that stems from the inevitable loss of information that accompanies scalar evaluation. Programs are complex combinatorial entities and program execution is

© Springer International Publishing Switzerland 2016
K. Krawiec, *Behavioral Program Synthesis with Genetic Programming*,
Studies in Computational Intelligence 618,
DOI: 10.1007/978-3-319-27565-9_2

a nontrivial process. Yet, in conventional GP all what is left of that process is
a single number, i.e. the number of failed tests for discrete domains (1.7) or
a continuous error in the case of regression problems (e.g., (1.8)).

main
claim
of this
book

The main tenet of this book is that the conventional scalar evaluation *de-
nies a search algorithm access to detailed behavioral characteristics of a
program*, while *that information could help to drive the search process more
efficiently*. This observation can be alternatively phrased using the message-
passing metaphor typical for information theory. A search algorithm and an
evaluation function can be likened to two parties that exchange messages.
The message the algorithm sends to the evaluation function encodes the candidate solution to be evaluated. In response, the algorithm receives a return
message – the evaluation. In a sense, the evaluation function *compresses*
a candidate solution into its evaluation. And compressing all information
about program behavior into a single number is inevitably lossy.

To illustrate the chasm between the richness of program behaviors and
the paucity of scalar evaluations, consider again synthesis of Boolean programs. Assume for the sake of argument that we identify program behavior
with the combination of outputs it produces for all tests (as in semantic
GP, Chap. 5). There are 2^{2^k} such behaviors of k-ary Boolean programs.
For the 11-bit multiplexer example considered in Sect. 1.4), this is 2^{2048}.
On the other hand, the objective function for this problem assumes only
$2^{11} + 1 = 2049$ distinct values (0 to 2048 inclusive), i.e. can be represented
with 11 bits (excluding the correct program). To believe that a search algorithm can efficiently search a space of 2^{2048} behaviors by obtaining 11
bits of information for each visited candidate solution is very optimistic, if
not naive – and recall that we consider the Boolean domain, arguably the
simplest one. It is hard to resist quoting the classic reflection on the size of
the Universe:

> Gigantic multiplied by colossal multiplied by staggeringly huge is
> the sort of concept we're trying to get across here. [1]

evaluation
bottleneck

The existence of this *evaluation bottleneck* is confirmed by practice. For
instance, even though contemporary GP algorithms manage to synthesize
11-bit multiplexers, the parity problem with the same number of variables
is for them very difficult and hardly ever gets solved.

2.2 Consequences of evaluation bottleneck

In this section, we discuss the implications of scalar evaluation that origi-
nate in the aggregation of outcomes of multiple interactions of a program
with the tests. Such aggregative functions prevail in generative program

synthesis, but have been also intensely studied in *test-based problems* [16, 24] (Sect. 4.1). A function that counts the failed tests (2.1) is the most common, additive representative of that class. The error functions commonly used in symbolic regression (like MSE (1.8) or Mean Absolute Deviation, MAD) also belong to this class, as does the less common multiplicative aggregations.

2.2.1 Discreteness and loss of gradient

Many evaluation functions used in GP are by nature *discrete*. The objective function f_o that counts the failed tests is quite an obvious example here (2.1): f_o can assume only $|T| + 1$ unique values. When candidate programs pass the same number of tests, f_o fails to provide a search gradient. This is particularly likely when T is small, which is often practiced in GP to reduce computational cost. However, having many tests does not necessarily solve this problem either, because programs in the population improve and with time it becomes more likely for them to pass the same number of tests. Also, in some domains and for some problems, achieving certain values of f_o is more likely than others; for instance the parity problem (see Chap. 10 for definition) is notorious for many programs having f_o that is a power of two. In such situations, the evaluation function (and consequently a selection operator) ceases to differentiate candidate solutions, the search gradient is lost, and search becomes purely random.

loss of gradient

Increasing the number of tests is not necessarily beneficial for yet another reason. Consider a set T of $\mathcal{T} = 10\,000$ tests and two program synthesis tasks defined for that problem, with evaluation functions based on randomly drawn subsets $T_1, T_2 \subset \mathcal{T}$ of, respectively, 100 and 1 000 tests. T_2 is ten times larger, so the values of an evaluation function that drives the corresponding search process are more *precise*, i.e. estimates better the true performance. But in the presence of a rugged fitness landscape (Sect. 1.4), better precision may be insignificant, and an evaluation function based on T_1 may perform equally well.

The problem of discreteness in principle does not apply to evaluation functions that aggregate continuous interaction outcomes (e.g., (1.8)), typically used in regression problems. However, due to the many-to-one mapping from programs to evaluations and limited number of tests, there are usually many programs that implement the same real-valued function. Such candidate solutions will receive identical evaluation and become indistinguishable, or worse, the presence of rounding errors will render one of them slightly better, leading to an unintended search bias.

2.2.2 Compensation

Another consequence of adopting an aggregative objective function like the objective function f_o is *compensation*: all programs that pass the same number of tests are considered equally fit, no matter *which* particular tests they pass. In other words, such a function is *symmetric* with respect to tests[1]. This characteristic ignores the fact that some tests can be *inherently* more difficult than others: in a given programming language $\mathcal{P}$, more programs may exist that pass a test t_1 than programs that pass some some other test t_2. Following [69], we can formalize the difficulty of a test $t = (in, out) \in \mathcal{T}$ for a programming language $\mathcal{P}$ as

$$diff(t) = diff((in, out)) = \Pr(p \in \mathcal{P} : p(in) \neq out). \qquad (2.2)$$

objective
test
difficulty

subjective
test
difficulty

We further refer to this quantity as to *objective test difficulty*. $diff(t)$ is thus the probability that a program randomly picked from $\mathcal{P}$ passes test t. Alternatively, one might define *subjective test difficulty*, i.e. the probability of passing a test by a program synthesized by a given search algorithm. Nevertheless, objective test difficulty $diff(t)$ is more universal and can be estimated even if $\mathcal{P}$ is large or infinite [69].

It is interesting to realize that encountering in the real world a problem with *diff* varying across tests in $\mathcal{T}$ is actually more likely than all tests being equally difficult. This is evidenced not only in program synthesis, but also in test-based problems (Sect. 4.1), e.g., in games [176].

2.2.3 Biased search

Discreteness and compensation are inherent properties of aggregative evaluation functions. More subtle consequences of scalar evaluation come to light only when considering the entirety of the program synthesis process. In the following, we demonstrate the interplay of characteristics of search operators with an objective function. To that end, we employ the formalism of *outcome vector*, i.e. the vector of the outcomes of program's interactions with particular tests:

outcome
vector

$$o_T(p) = ([p(in) = out])_{(in, out) \in T}, \qquad (2.3)$$

where $[\,]$ is the Iverson bracket, defined for a logical predicate α as

$$[\alpha] = \begin{cases} 1, & \text{if } \alpha \text{ is true} \\ 0, & \text{otherwise} \end{cases} \qquad (2.4)$$

The ith element of $o_T(p)$ is 1 if p passes the ith test in T, and 0 otherwise. As signaled earlier, T needs to be a list, not a set, for o_T to be well-defined.

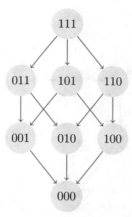

Fig. 2.1: The lattice of outcome vectors for an abstract problem featuring three tests. Value 1 denotes passing a given test, 0 – failing it. Blue arrows mark the dominance relation between outcome vectors.

Notably, an outcome vector is the first non-trivial *behavioral descriptor* the reader encounters in this book.

For a given T, the set of all possible outcome vectors forms a special case of partially ordered set (*poset*), a *lattice*. The lattice has $2^{|T|}$ nodes; Figure 2.1 presents an example for three tests. The top element in the lattice corresponds to the target of search (1.4) and groups optimal solutions, i.e. programs that pass all tests and thus achieve $f_o(p) = 0$. The bottom element correspond to the worst solutions ($f_o(p) = |T|$). The outcome vectors in the intermediate layers have evaluation varying from $|T| - 1$ (second layer from the bottom) to 1 (second layer from the top). A generate-and-test program synthesis method like GP will usually start with the programs occupying middle levels in the lattice, and progress toward the top node.

Any program interacting with the tests in T can be unambiguously assigned to one and only one node in this lattice based on its outcome vector. Multiple programs may occupy the same node in the lattice due to the mapping from programs to their behaviors being many-to-one (Sect. 1.4). For most programming languages $\mathcal{P}$, some behaviors of programs in $\mathcal{P}$ are more common than others. As a result, the distribution of programs over lattice nodes is usually highly non-uniform (see [106, Chap. 7] for an example for the Boolean domain).

The arcs in Fig. 2.1 illustrate the *dominance relation* between outcome vectors. An arc connects a node o_1 to o_2 if and only if every test passed in o_1 is also passed in o_2 *and* there is exactly one test in o_2 that is not passed in o_1. However, these arrows have little to do with possible moves of a search

---

[1] Unless it differentiates their importance in some way (e.g., by weighing)

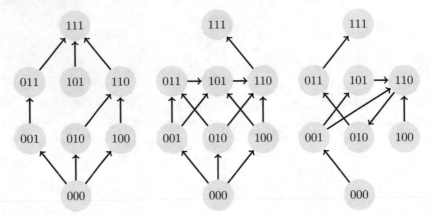

Fig. 2.2: Transition graphs on outcome vectors for three abstract problems, each featuring three tests. Contrary to Fig. 2.1 where arrows mean dominance, here they mark the changes of outcome vectors (behavioral, phenotypic changes) resulting from modifications applied to programs by a search algorithm (genotypic changes).

algorithm. How a given program synthesis algorithm traverses such a lattice is in general unrelated to the dominance relation. The reason for that is the highly non-local genotype-phenotype mapping discussed in Sect. 1.4. A search operator may not be able to improve a current candidate program (with an outcome vector in o_2) by augmenting it with the capability of solving yet another test while preserving that capability for the already passed tests (so that it ends up in o_1). On the other hand, the same search operator may be capable of flipping *multiple* elements of the outcome vector of a given program in a single application, moving a program several levels up, down, or sideways in the lattice.

Therefore, in the figures that follow we drop the dominance edges and represent only the transitions between outcome vectors as induced by a given search operator. Though the spatial arrangement of nodes remains the same, these posets are not lattices anymore and will be referred to as *transition graphs*. Similar graphs have been previously considered in [60], however without taking into account the spatial arrangement of nodes resulting from stratification of evaluation.

transition
graph

Example 2.1. Consider a hypothetical task with such a transition graph shown in Fig. 2.2a. The arrows mark the possibility of transitions between outcome vectors. For instance, the arc from the node labeled 010 to the node labeled 110 indicates that there exists a program with the outcome vector 010 such that it can[2] be modified by the considered search operator so that its outcome vector changes to 110. Note that in general one should not expect the arrows to be mirrored: a search operator may be unable to revert the effects of its application.

It does not take long to realize that the problem shown in Fig. 2.2a is easy: the transitions are aligned along the gradient of the objective function f_o, so even a straightforward search algorithm (e.g., greedy search) should easily traverse the path from 000 to 111. Now consider the problem shown in Fig. 2.2b: here, the transition from 101 to 110 is not accompanied by an improvement (f_o remains to be 1). Because the evaluation function imposes only a vertical gradient in the graph, if a search algorithm that accepts only strict improvements happens to visit 101, it will get stuck there. This does not seem to be much of a problem for stochastic search algorithms like GP, which could still move from 101 to 110 by pure chance. Consider however the problem shown in Fig. 2.2c. Once a search process reaches the combination 110, further progress can be made only by moving to 010, which implies the deterioration of f_o from 1 to 2. Only search algorithms capable of accepting such deterioration will be able to overcome this trap. Again, stochasticity of GP will occasionally permit that, but the likelihood of such an event may vary, and such transition graphs are in general harder to traverse than those shown in Figs. 2.2a and 2.2b.   ∎

To an extent, the above graphs are related to *fitness landscapes* [194]. Fitness landscapes visualize a scalar evaluation function stretching over a solution space $\mathcal{P}$ arranged with respect to the neighborhood induced by a search operator. The case in Fig. 2.2a is characteristic for a unimodal fitness landscape devoid of *plateaus*, i.e. neighboring solutions with the same fitness. The case in Fig. 2.2b reveals presence of some plateaus (e.g., (011,101,110)), and the one in Fig. 2.2c features *traps* (110,101) and can be thus characterized as *deceptive*.

fitness landscape

This is however where the analogy with fitness landscapes ends. A point on a fitness landscape corresponds to one candidate solution. The nodes in Fig. 2.2 correspond to behavioral *equivalence classes* induced by outcome vectors. By this token, they provide more detailed behavioral information and can prospectively allow for identifying more nuanced structures and challenges to them. Recall however that a single node gathers here *multiple* programs (some of them possibly syntactically different) that have the same behavior. That behavior is the only thing they have in common: some of those programs may be able to undergo certain modifications of a search operator, while some not. In this sense, these graphs paint an overly

behavioral equivalence class

---

[2] 'Can', because the outcome of an application of a *stochastic* search operator to a given program is non-deterministic.

optimistic picture about the possible traversals in the space of outcome vectors.[3]

2.3 Experimental demonstration

The examples in Fig. 2.2 are intentionally simple and abstract from a specific domain. In practice, the behavior of a given search algorithm on a given problem is more complex. The transitions between outcome vectors, rather than being possible or impossible, become more or less *likely* due to various biases of search operators. In GP, another reason for that is the stochastic nature of search operators. It is thus justified to ask: are the characteristics discussed above and exemplified in Fig. 2.2 purely hypothetical or do they occur in real-world domains? To answer this question, we experimentally construct analogous graphs for selected real-world problems.

We consider the Boolean domain and programs composed of instructions $\{and, or, nand, nor\}$. To construct a graph, we first generate a sample of $10\,000\,000$ random programs of depth up to 8 using the conventional ramped half-and-half method [79]. For each generated program p, we mutate it, obtaining a program p', calculate the outcome vectors $o(p)$ and $o(p')$, and collect the statistics of transitions between outcome vectors over the entire sample. We employ the subtree-replacing mutation operator commonly employed in GP [148], which uses ramped half-and-half with depth limit 8 to generate random subtrees.

Figure 2.3 visualizes the resulting transition graph for the Boolean 3-bit multiplexer problem (MUX3). In this problem, program input comprises three variables: the first one serves as the 'address line' that decides which of the remaining two input variables should be passed to the output – this is the desired behavior of a correct program. Given three tests, the objective function (1.7) varies from 0 to 8 inclusive. Even though MUX3 is trivial for contemporary program synthesis methods, this small domain is already quite rich in terms of program behavior: there are $2^8 = 256$ possible outcome vectors (and thus nodes in the lattice), and the central layer of the lattice features $\binom{8}{4} = 70$ nodes. For this reason, Fig. 2.3 shows only the three top layers of the lattice, i.e. the outcome vectors that correspond to evaluation 0 (optimal solutions, one node in the top layer), 1 ($\binom{8}{1} = 8$ nodes in the second layer) and 2 ($\binom{8}{2} = 28$ nodes in the third layer). The topmost layers are critical for the success of program synthesis, as in practice search often gets stuck with one or two failing tests and therefore does not make

---

[3] This leads to an interesting observation concerning the nodes without improving outgoing arrows: although many programs that occupy them are syntactically different, they are all hard to improve.

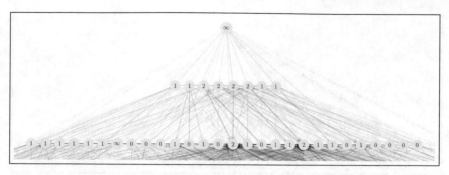

Fig. 2.3: The top three layers of the transition graph on outcome vectors for the MUX3 problem. See [82] for the complete figure.

further progress. The reader is invited to download the image of the entire lattice provided online at [82] to conveniently zoom into details.

The spatial arrangement of the nodes in Fig. 2.3 is analogous to that in Fig. 2.2. This time however the widths of the arrows vary and reflect the estimated probability of transitions. Given a node o, the widths of the outgoing arcs are proportional to the probabilities of mutation moving a program with outcome vector o to successor nodes. For clarity, we do not draw the arcs corresponding to neutral mutations (which would start and end in the same node) and arcs that account for less than 1 percent of outgoing transitions.

Because we are not going to inspect specific outcome vectors, we do not label the nodes with them but with the estimated log-odds of improvement. For a given node, the label is

$$- \left\lfloor \log_{10} \frac{n^-}{n^+} \right\rfloor, \tag{2.5}$$

where n^- is the number of non-improving moves outgoing from that node, and n^+ is the number of improving moves outgoing from the node. For instance, a label '2' indicates that the odds of non-improving moves to the improving ones is of the order of $10^2 : 1$, i.e. $100 : 1$. The ∞ symbol labels the nodes for which no improving move has been found in the sample.

The log-odds clearly increase with improving evaluation f_o, i.e. with the decreasing number of failed tests (1.7). As observed in typical GP runs, with closing to the target (corresponding to the top node of the graph), it becomes harder to make improving moves. This trend becomes even more evident when confronted with the lower layers not shown in print. Nevertheless, the chances of reaching the optimal solution in this example are clearly not negligible, as evidenced by the large number of the arcs incoming to the top node of the lattice. This is due to simplicity of MUX3.

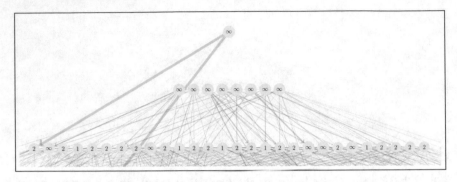

Fig. 2.4: The top three layers of the transition graph on outcome vectors for the PAR3 problem. See [82] for the complete figure.

In Fig. 2.4, we present an analogous graph for the much harder PAR3 problem, where a correct program should return *true* if and only if the number of input variables that have the value *true* is even. The graph, generated using the same instruction set and identical settings as for MUX3, features a substantially different structure of arcs. The log-odds of the orders between 1 and 2 occur already one layer lower than for MUX3. In the second layer from the top, all nodes are marked by ∞ and all the outgoing arrows point downwards: no program with these outcomes vectors in the considered sample underwent an improving mutation.

As a side remark, it is interesting to note that, for binary output domains ($|\mathcal{O}| = 2$), there is one-to-one correspondence between the outputs of a program (*true, false*) and test outcomes $(0, 1)$. Therefore, other things being equal, Figs. 2.3 and 2.4 present, as a matter of fact *the same* transition graph, however with node locations permuted spatially, so that layers group the nodes that have the same evaluation in a given problem. More generally, this graph is common for *all* 3-bit Boolean problems.

2.4 Discussion

The conclusion following from Figs. 2.3 and 2.4 is that *programs' odds for being improved vary by orders of magnitude*. Moreover, the outcome vectors that offer the highest chance of further improvement are not necessarily the easiest ones to attain. For instance, the nodes labeled with 1 in the second layer of Fig. 2.3 (10 : 1 odds of non-improving transitions) do not seem to feature more incoming arcs than the less attractive nodes, i.e. those labeled with 2 (odds 100 : 1).

Scalar evaluation performed by the objective function does not reveal such differences. Programs with outcome vectors in the same graph layer receive the same evaluation and render themselves indistinguishable. Conventional evaluation functions combined with biases of search operators favor certain paths of accretions of *skills*, meant here as capability to pass particular tests. As a result of that bias, certain outcome vectors become particularly easy to attain. If a given outcome vector does not offer an opportunity for further improvement, it forms a counterpart to the conventional notion of *local optimum*. In a longer run, the programs with easy-to-attain outcome vectors tend to prevail in a population, causing it to converge prematurely.

skill

local optimum

premature convergence

Stochastic approaches to program synthesis like GP are *in principle* resistant to premature convergence, because a well-designed stochastic search algorithm visits in the limit *all* points in the reachable search space. However, guarantees 'in the limit' are of little use in practice. In general, GP does not scale well with the growing number of tests and complexity of a synthesis task. It seems thus justified to address the above issues by *making the search algorithm more aware about the differences in program behavior*. This is the revolving theme of this book, and the following chapters review and propose practical means to this end.

Another upshot is that it is not the *number* of tests passed, but the particular *combination* of them that may be critical for a search algorithm to solve a program synthesis task. The interplay of scalar evaluation function with the characteristics of search operators may lead to a situation in which a candidate solutions may need to first master some skills before attempting to master others. This intuition gave rise to some of the GP extensions presented in the next chapters. It was also present in the studies on test-based problems and coevolutionary algorithms [16, 24, 149].

2.5 Related concepts

The limited capability of scalar evaluation functions has been identified in many branches of computational intelligence and beyond. Within evolutionary computation, this problem is usually tackled with techniques. Implicit fitness sharing (Sect. 4.2) and novelty search (Sect. 9.10) are among them. Other approaches include niching [118] and island models [191].

diversity maintenance

The realization of the limited 'informativeness' of the evaluation function has also surfaced beyond evolutionary computation. Deep learning, responsible for the recent progress in the field of artificial neural networks is, at least to some extent, motivated by this observation. Let us quote here one of the most often cited papers in that domain:

deep learning

> Training a deep network to directly optimize only the supervised objective of interest (for example the log probability of correct classification) by gradient descent, starting from random initialized

parameters, does not work very well. (...) What appears to be key
is using an additional unsupervised criterion to guide the learning
at each layer. [185]

The 'additional unsupervised criterion' is intended to provide an extra (or
alternative) guidance for a learning algorithm (an analog to search algorithm in GP). In the cited work, it helps to learn higher-level representations of input data while preserving their key features. Interestingly, this
is analogous to the approach we proposed in [100] and present in Chap. 6,
where evolving programs are promoted for preserving the relevant information contained in tests at intermediate stages of program execution (which
may be likened to hidden layers in a multilayer neural network). Crucially,
in doing so we do not actually specify *what* a program should produce at an
intermediate execution state. In this sense, we guide thus a search process
in an unsupervised way, as in the above quote.

Last but not least, it may be illuminating here to adopt for a moment
the biological perspective, the source of inspiration for EC and GP. The
possibility of an evaluation function failing to efficiently drive a search
process should not come as a surprise, because fitness in biology is not
meant to *drive* anything: it is a figment of the conceptual framework in
which we phrase the evolution in Nature. Fitness reflects the reproductive
success of a given individual (see, e.g., Sect. 2.2 in [139]), and as such is
known only a posteriori, once the parents have been selected (for *absolute*
fitness) or once the statistics on the descendants of the current individuals
are known (for *relative fitness*). And although it summarizes the entirety of
individual's skills that are relevant for gaining an evolutionary advantage,
it cannot be reverse-engineered: it is virtually impossible to guess from a
fitness value *which* skills made a given individual successful.

This observation is related to the fact that the natural evolution does not
optimize for anything; rather than that, it opportunistically improves the
skills (and combinations thereof) that increase the odds of survival (or
comes up with completely new skills). For this reason, the EC paradigm
that should be considered the closest to the biological archetype is[4] that of
coevolutionary algorithms, where individuals (or species, or individuals and
their environments) impose selective pressure on each other, rather than
such a pressure originating in some external objective function (Sect. 4.1).

Another biological aspect worth mentioning here concerns the relationship
between the aggregative nature of conventional evaluation functions used
in EC and *gradualism*, the cornerstone of classical Darwinism (see also the
comments on *accretion* in Sect. 1.5.3). As gradualism posits that progress
in natural evolution is slow and steady, so passing each new test gives an

---

[4] Apart from the open-ended evolution, which is however less common for EC
and more for artificial life.

individual in EC only a minute evolutionary advantage (unless the number of tests is low, which is of no interest here). However, even if indeed the biological evolution was gradual in this sense (though it has been questioned many times), we posit that enforcing an analogous approach in EC is not necessarily effective. Humans rarely solve problems gradually, case by case. Instead, they identify regularities and perform conceptual leaps via inductive or deductive reasoning. Interestingly, the temporal dynamics of this process can be likened to *punctuated equilibria*, i.e. long periods of no improvement interlaced with sudden rises in fitness, observed both in natural evolution and EC [186].

punctuated equilibria

2.6 Summary and the main postulate

This chapter gathered the evidence that *evaluation bottleneck* resulting from reliance on conventional scalar evaluation functions has detrimental consequences, including loss of search gradient and premature convergence. Arguably, there are domains where an evaluation function is by definition 'opaque' and makes this bottleneck inevitable. For instance, in Black Box Optimization, the value of the evaluation function is the only information on candidate solutions available to a search algorithm. Similarly, *hyperheuristics* [18] usually observe the *domain barrier* that parts the search algorithm from a problem.

hyperheuristics
domain barrier

However, it might be the case that the need of such separation is more an exception than a rule when considering the whole gamut of problems we tackle with metaheuristics. In many domains, there are no principal reasons to conceal the details of evaluation. This is particularly true for program synthesis, where an act of evaluating a candidate program produces detailed information that can be potentially exploited. There are at least two levels of detail involved in there. Firstly, a program interacts with *multiple* tests. Secondly, a program's confrontation with a single test involves executing *multiple instructions*, each of them having possibly nontrivial effects.

The main claim of this book is that *the habit of driving search using an objective function alone (especially a scalar one) cripples the performance of search algorithms* in many domains (in particular in generative program synthesis), and it *should be abandoned in favor of more sophisticated and more informative* search drivers. By providing search algorithms with more detailed information on program behavior we hope to broaden the evaluation bottleneck and improve their performance.

main claim

It may be worth mentioning that evaluation bottleneck is also awkward in architectural terms, i.e. when looking at a program synthesis system as a network of interconnected components, shown for GP in Fig. 1.1. Why compress evaluation outcomes in one component (evaluation function) and then force another component (e.g., selection operator) to 'reverse-engineer' them? There are no other reasons for this other than the convention inherited from metaheuristic optimization and excessive adherence to – not necessarily accurate, as we argued in Sect. 2.5 – evolutionary metaphor.

In the following chapters, we present two mutually non-exclusive avenues toward that goal. The first, represented in this book by semantic genetic programming (Chap. 5), consists of designing search operators that produce offspring in a more 'directed' way. The other approach is exercised here more extensively and aims at alternative multifaceted characterizations of program behavior. This philosophy is embodied by implicit fitness sharing and related methods (Chap. 4), which analyze convergence of execution traces (Chap. 6), and pattern-guided program synthesis (Chap. 7). The next chapter introduces the common formal framework that unifies those approaches and facilitates their presentation.

3

The framework of behavioral program synthesis

In the previous chapter, we identified evaluation bottleneck, and showed that the information lost in aggregation of interaction outcomes can be essential for the success of a program synthesis process. In this chapter, we set out to present *behavioral program synthesis*, a new paradigm of program synthesis which, in a sense, puts program behavior before its source code.

behavioral program synthesis

Behavioral program synthesis does not offer any specific algorithm or even an algorithm template. In tune with [168] (see also Sect. 11.2), we see it rather as a consistent toolbox of interlinked formalisms that can be used in different contexts. The key concepts of behavioral program synthesis are program trace and execution record, which facilitate design of better-informed program synthesis algorithms presented in the subsequent chapters.

3.1 Program traces and execution records

The list of shortcomings of conventional objective functions presented in the previous chapter suggests that it may be worth to look for alternative means of characterizing program performance. Such means should better inform a search algorithm about other aspects of program behavior and so broaden the bottleneck of scalar evaluation.

As a matter of fact, several existing extensions of the traditional GP paradigm build upon this observation. For instance, program semantics in GP is a vector of program outputs for particular tests (Chap. 5) and thus provides more information about program behavior than the conventional scalar evaluation. Behavioral descriptors like this are tailored to the needs of a particular approach. A program semantic in GP holds program output for every test, because this is the information required by and sufficient for semantic-aware search operators.

© Springer International Publishing Switzerland 2016
K. Krawiec, *Behavioral Program Synthesis with Genetic Programming*,
Studies in Computational Intelligence 618,
DOI: 10.1007/978-3-319-27565-9_3

Contrary to this model, we propose that evaluation should provide a *complete account* of program behavior, and to leave it up to the other components of a search algorithm to decide which pieces of that information to use. The means for this is the *execution record*, which reports the course of program execution for all tests available to evaluation function. To present this formalism, we must delve deeper into program execution process.

We assume that programs are run in a certain *execution environment* (*environment* for short). The execution environment comprises all components that can be affected by program and such that their state is well-defined between execution of consecutive instructions. In conventional hardware architectures, environment would include the accessible memory (RAM and processor registers) and all stateful I/O devices. In program synthesis practiced with GP, the environment is typically more humble. In tree-based GP, which typically implements functional programming and is thus free of side effects, the environment is simply the value returned by the currently executed node of an expression tree. In linear GP [14], statements affect registers and the environment would comprise them all. In the PushGP system [170], execution environment includes code stacks and data stacks [101].

Regardless of the programming paradigm and program representation, at any given moment of program execution, the environment is in a certain *execution state* (*state* for short). Program execution consists of an iterative application of *program interpreter* (*interpreter* for short) to the states of an environment. Execution of consecutive instructions of a program generates a sequence of states, called *program execution trace* (*trace* for short). A trace of a program p applied to an input in can be denoted as

$$p^0(in), p^1(in), p^2(in), p^3(in), \ldots \qquad (3.1)$$

Here, $p^i(in)$ stands for the state of the environment after executing i steps of program p. In particular, $p^0(in)$ denotes the *initial execution state*, which directly depends on the input in submitted to the program. How in specifically determines $p^0(in)$ depends on the particular programming paradigm. For instance, in tree-based GP, in determines the state of all non-constant terminal nodes in a program tree. In Push, in determines the data placed on the working stacks.

If, as we assumed in Sect. 1.1, a program halts for a given input, its trace is finite, and the last state determines the output of a program. The particular interpretation of 'determines' depends on the programming paradigm. For tree-based GP, the last state is the value returned by the root node of a program tree, because this is where the data flow terminates. In linear GP and PushGP, the definition of program output is more arbitrary; typically, whatever a program has left on the top of one of the data stacks (in PushGP) or in a designated register (in linear GP) is interpreted as its output.

A program trace of a halting program applied to an input *in* can be thus presented in a more complete way as:

$$in \to p^0(in), p^1(in), p^2(in), p^3(in), \ldots, p^\$(in) \to out \qquad (3.2)$$

where the mapping $in \to p^0(in)$ determines the initial execution state based on the input *in*, and $p^\$(in) \to p(in)$ determines program output *out* based on the final execution state $p^\$(in)$. The dollar symbol marks the last state of trace, whatever its actual length is.

In general there is no one-to-one correspondence between the instructions and the states in a trace of a program. A trace can be longer or shorter than the program it originated from; the former may be due to loops or recursion, and the latter due to conditional statements. Traces are thus strictly behavioral and should not be confused with execution paths in ASTs or block diagrams.

With program trace defined, we are ready to formalize an execution record. Given a list of m tests T, the *execution record* of a program p is the list of traces obtained by applying p to every test $(in_i, out_i) \in T$, i.e.   execution record

$$
\begin{aligned}
&p^0(in_1), p^1(in_1), p^2(in_1), \ldots, p^\$(in_1) \\
&p^0(in_2), p^1(in_2), \ldots, p^\$(in_2) \\
&\ldots \\
&p^0(in_m), p^1(in_m), p^2(in_m), p^3(in_m), p^4(in_m), \ldots, p^\$(in_m)
\end{aligned}
\qquad (3.3)
$$

where $p^i(in_j)$ is the ith element of the trace generated by applying p to the jth test in T. As usually, we assume that p halts for all inputs in T. Observe that trace lengths may vary with tests.

Let us emphasize that an execution record does not hold *any* traces, but the traces that are implicitly linked to each other by the underlying program. On the other hand, the execution record is not bound to any specific program synthesis task, and thus agnostic about correctness predicate, desired outputs (target), and objective function. It depends only on the program and the list of inputs that come from the tests of consideration T.

An execution record forms the full account of program behavior for the inputs in T. There is nothing about program behavior that could not be deduced from it[1]. It is trivial to show that execution record allows computing, amongst other things, the conventional objective function. Inspecting the final execution states in every trace in (3.3) allows us to rephrase f_o (Eqs. 1.7 and 4.2) as

$$f_o(p) = |\{(in, out) \in T : p^\$(in) \neq out\}|. \qquad (3.4)$$

---

[1] Unless we consider non-functional properties not reflected in program traces like execution time.

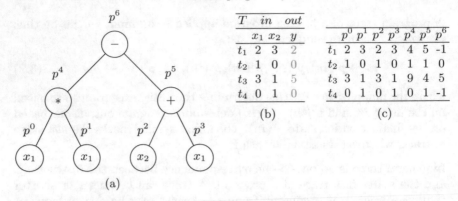

T	in		out
	x_1	x_2	y
t_1	2	3	2
t_2	1	0	0
t_3	3	1	5
t_4	0	1	0

(b)

	p^0	p^1	p^2	p^3	p^4	p^5	p^6
t_1	2	3	2	3	4	5	-1
t_2	1	0	1	0	1	1	0
t_3	3	1	3	1	9	4	5
t_4	0	1	0	1	0	1	-1

(c)

(a)

Fig. 3.1: An exemplary program (a), a list of tests T, each comprising two input variables x_1 and x_2 (b), and the execution record resulting from applying the program to the tests in T (c). The annotations atop the instruction nodes in (a) mark the corresponding columns in the execution record ((c). The desired output in T is grayed out to emphasize that execution record is oblivious to it. See Fig. 3.2 for different illustration of (b) and (c).

The confrontation of f_o with the richness of execution record makes us realize again that the conventional evaluation function provides only a very crude summary of program behavior. In the following chapters, we base several alternative measures on execution record and use them to drive search.

3.2 Realization of execution record

The execution record is intentionally generic to allow embracing different program representations. When applied to a particular genre of GP, it needs to be tailored to its specifics, which may involve certain simplifications.

Execution of a single instruction of a program is usually highly local in terms of effects. In imperative programming languages it may modify a single variable, leaving all other components of the execution environment intact. In functional programming, it affects only the value returned by the currently executed function. The following example illustrates the consequences of this characteristics.

Example 3.1. Figure 3.1 presents a tree-GP program, a list of four tests, and the corresponding execution record. The input part of every test comprises two input variables x_1 and x_2. The program in question is an expression (does not feature loops nor conditional statements), thus the states in the execution record correspond one-to-one to program instructions. Because instructions have no side effects here, the ordering of their execution is only partially determined by the structure of program tree. The only constraint is that the arguments of an instruction have to be computed prior to its execution; the order of arguments' execution is irrelevant. There are thus many ways in which such functional programs can be executed. In Fig. 3.1, we assume execution order from bottom to top, and from left to right.[2]

In this example, execution states are equivalent to values returned by instructions. There is no other information that a state needs to store in order to capture the effects of computations conducted within the corresponding subtree.

Note that an execution record contains also the input parts of the tests, as input data get reflected in the initial states p^0 of traces. In Fig. 3.1, the four initial states in every trace hold the copies of the input variables x_1 and x_2. ∎

Example 3.1 presented the technical details of constructing an execution record in tree-based GP. This process will vary for other GP paradigms; the reader is referred to [101] for analogous explanations for PushGP [170]. Nevertheless, in all genres of GP, subsequent execution states are often redundant with respect to their predecessor: for instance in PushGP, an single instruction will usually affect only top elements of selected stacks. Therefore, execution states considered in this book will reflect only the most recent computation – typically the outcome of the most recent instruction. Such 'differential' implementation of an execution record has also the technical advantage of reducing the memory footprint, while still reflecting the entirety of effects of computation.

differential execution record

Traces in a given execution record may vary in length (3.3) as a result of conditional statements, loops, or recursion. However, some of the methods presented in this book require traces to be *aligned* so that the states in traces correspond to each other, i.e. reflect the environment at the same stage of execution for particular inputs. To meet this requirement, in the following we consider only *expressions*, i.e. non-recursive programs that are free of conditional statements and loops[3]. For expressions, execution states correspond one-to-one to program instructions, and an execution record can be represented as a two-dimensional array (rather than a list of lists (3.3) where rows correspond to tests and columns to instructions:

alignment of execution traces

---

[2] This observation suggests alternative ways of defining traces: not as linear structures, but for instance as trees that reflect the structure of a program. However, for clarity and consistency with the past work, we adopt the list-based representation for traces.

[3] This assumption is however not particularly severe, given that contemporary work in GP focuses on such programs anyway

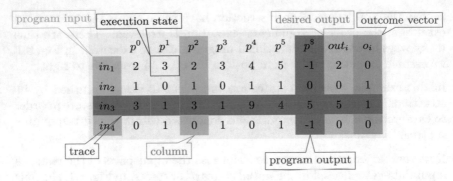

Fig. 3.2: Visual glossary of the terms related to execution record, for the execution record and tests from Fig. 3.1. The execution record comprises the columns from p^0 to $p^\$$ inclusive.

$$
\begin{array}{ccccc}
p^0(in_1) & p^1(in_1) & p^2(in_1) & \ldots & p^l(in_1) \\
p^0(in_2) & p^1(in_2) & p^2(in_2) & \ldots & p^l(in_2) \\
\vdots & \vdots & \vdots & \vdots & \vdots \\
p^0(in_m) & p^1(in_m) & p^2(in_m) & \ldots & p^l(in_m)
\end{array}
, \qquad (3.5)
$$

where l denotes program length.

When it comes to implementation, an execution record can be acquired in at least two ways. The most obvious approach is to 'instrument' a program, i.e. place *traps* (breakpoints) in its source code, and make a snapshot of the environment every time execution reaches a trap. The alternative method requires access to the source code of interpreter that runs the programs. Interpreter code usually contains a loop that realizes the actual control flow by fetching program instructions and executing them one by one. By modifying that loop, one can halt the execution after every such a cycle, and intercept the current execution state. As a matter of fact, this functionality is built-in into quite many software and hardware platforms in order to support debugging.[4,5]

---

[4] A hardware-level implementation is a trap flag, present in virtually every CPU. When set, it causes CPU to halt after executing each instruction. A software-level example is the `trace()` function in the R programming language that allows setting a callback function to be invoked after every execution cycle (see www.r-project.org).

[5] Note that if a GP interpreter is stateless, execution state needs to include also some notion of *instruction pointer* so that it is known from where to resume execution. In some GP paradigms, a form of instruction pointer is already built-in. In PushGP [170], the code yet to be executed is stored on the CODE stack. On the other hand, tree-GP interpreters are typically implemented via recursion and have no explicit instruction pointers.

3.3 Summary

This chapter concludes the introductory part of this volume. To summarize the main concepts of behavioral program synthesis, in Fig. 3.2 we present a 'visual glossary' for an exemplary program and execution record. This presentation emphasises that there are at least five qualitatively different conceptual levels for characterizing program behavior, which correspond to various *behavioral descriptors* considered throughout this book. We list them here along with the data types they represent, assuming m tests and domain $(\mathcal{I}, \mathcal{O})$:

behavioral descriptor

1. Program correctness, $\mathbb{B}$,

2. Scalar evaluation, $\mathbb{R}$,

3. Outcome vector (a row in an interaction matrix), $\mathbb{B}^m$,

4. Output vector (*program semantics* in Chap. 5), $\mathcal{O}^m$,

5. Execution record, $\mathcal{T}^{m \times n}$, where n is the length of a program (and the corresponding aligned execution record), and $\mathcal{T}$ is the type of trace elements (or a placeholder for multiple types if required).

The order of these behavioral descriptors reflects the increasing amount of behavioral 'capacity', i.e. amount of information conveyed regarding program behavior. For this reason, the vector-based behavioral descriptors (3 and 4) are placed in between scalar evaluation (1) and execution record (which is technically a matrix (3.5)). Program semantics is richer than an outcome vector as it holds the actual output of a program, while an outcome vector reveals only if particular tests have been passed or failed. Note that 1, 2 and 3 are problem-specific (depend on the target), while 4 and 5 are not (do not involve the target).

In subsequent chapters we demonstrate how execution records broaden the evaluation bottleneck and facilitate the opening of the 'black-box' of evaluation. However, their potential can be leveraged only if a search algorithm can exploit it. The question to be answered is thus: how can the particular components of an iterative program synthesis algorithm benefit from the availability of execution record? We answer this question for two stages of evolutionary workflow: selection (Chaps. 4, 6 and 7) and search operators (Chaps. 5 and 8). In perspective, execution records will allow us to construct *search drivers*, the other core concept of this book to be formalized in Chap. 9. The order of the chapters reflects the increasing conceptual sophistication of methods and the amount of information they scrutinize in an execution record.

Behavioral assessment of test difficulty

As argued in Sect. 2.2.2, one of the vices of conventional scalar evaluation is symmetry: the same reward is granted for passing every test. Yet some tests can be objectively more difficult than others in the sense of (2.2), i.e. harder to pass by a randomly generated program. They may vary also with respect to subjective difficulty, i.e. particular program synthesis methods may find it more or less difficult to synthesize a program that passes a given test (cf. Sect. 2.2.3). Conventional evaluation function (1.7), by simply counting the failed tests, cannot address this aspect of program synthesis.

In theory, test difficulty can be obtained from domain knowledge or provided by a human expert. But human expertise and domain knowledge are not always available or affordable, not to mention the extra effort required in such scenarios.

In this chapter, we show how information on test difficulty can be conveniently acquired from an execution record and used to redefine an evaluation function. This idea materialized originally in GP with the advent of *implicit fitness sharing* [166], which we cover in Sect. 4.2. In subsequent sections, we present the conceptual progeny of that approach: the methods that scrutinize cosolvability of tests [94] (Sect. 4.3) and automatically derive objectives from interaction matrices [112, 95] (Sect. 4.4). Before presenting that material, we first introduce the test-based perspective on program synthesis, which comes in particularly handy for this kind of considerations.

4.1 Test-based problems

Evaluation in GP can be alternatively phrased as candidate programs engaging in *interactions* with tests. In that framing, the evaluation of a candidate program p depends on an *interaction function* $g : \mathcal{P} \times \mathcal{T} \to \{0, 1\}$, interaction function which is an indicator function of the set of tests passed by p, i.e.

$$g(p, t) = g(p, (in, out)) = [p(in) = out], \qquad (4.1)$$

© Springer International Publishing Switzerland 2016
K. Krawiec, *Behavioral Program Synthesis with Genetic Programming*,
Studies in Computational Intelligence 618,
DOI: 10.1007/978-3-319-27565-9_4

where [] is the Iverson bracket (2.4). The objective function f_o (1.7) can be then rewritten as

$$f_o(p) = \sum_{(in,out) \in T} g(p, (in, out)). \qquad (4.2)$$

For convenience, we will occasionally abuse notation and treat g as a logical predicate, i.e. write $g(p,t)$ when $g(p,t) = 1$ and $\neg g(p,t)$ when $g(p,t) = 0$.

The outcomes of interactions of all programs in a given population P with

interaction all tests from a given set T can be gathered in an *interaction matrix*
matrix

$$G = [g_{ij} = g(p_i, t_j) : p_i \in P, t_j \in T]. \qquad (4.3)$$

Note that the ith row of G is the outcome vector for p_i, i.e. $o_T(p_i)$ (2.3), or in other words the rightmost column of an execution record (Fig. 3.2).

Formalizing an evaluation function in terms of interactions is not par-
coevolut- ticularly common in GP and more typical for coevolutionary algorithms
ionary
algorithm (CoEAs, [149]), where it originated in *test-based problem* [16, 24]. In a test-
test- based problem, one seeks an element or a combination of elements from
based
problem solution space $\mathcal{S}$ that conforms a given *solution concept* [30, 149]. The ar-
guably simplest example of solution concept is *maximization of expected utility*, i.e. a candidate solution that maximizes the expected interaction outcome, i.e. $\arg\max_{s \in \mathcal{S}} \mathbb{E}_{t \in \mathcal{T}} g(s, t)$.

The number of tests in $\mathcal{T}$ is usually large or infinite, which makes it techni-
cally infeasible to elicit exact values of an evaluation function. This problem can be addressed by sampling the tests to be used for evaluation, which can be done in at least three ways. In the simplest scenario, a sample T is drawn once from $\mathcal{T}$ and remains fixed throughout a run of a method; this case resembles a typical GP setup the most. Alternatively, one may sample T from $\mathcal{T}$ repetitively, for instance in each generation [21]. The third way is to let the tests *coevolve* with the candidate solutions in a CoEA framework. Typically, candidate solutions and tests co-evolve in two sep-
arate populations $S \subset \mathcal{S}$ and $T \subset \mathcal{T}$, respectively, interacting with each other only for the purpose of evaluation. While the candidate solutions are, as usually, rewarded for *performing*, the tests are rewarded for *informing*, for instance for the number of pass-fail distinctions they make between the current candidate solutions. The underlying rationale is that a CoEA can autonomously induce a useful search gradient by assorting the tests, and find good solutions faster or more reliably and/or at a lower computational cost compared to using all tests from $\mathcal{T}$ (where feasible) or drawing them at random. Empirical evidence gathered in previous work on various test-based problems suggests that this is indeed possible, provided proper tuning of a CoEA [69, 175].

Though CoEAs are not explicitly used in the methods studied in this book, the test-based framework is convenient for capturing the diversity of behaviors in an evolving population. Crucially, it allows to juxtapose not only the

behaviors of programs, but also compare the characteristics of tests, which is the idea behind the approaches presented in next sections.

4.2 Implicit fitness sharing

Implicit fitness sharing (IFS) introduced by Smith et al. [166] and further explored in GP by McKay [124, 123] originates in the observation that difficulty of particular tests may vary. Let us reiterate after Sect. 2.2.2 that problems with uniform distribution of test difficulty are less common than problems where difficulty varies by tests, as the former is a special case of the latter. The conventional objective function (1.7) is oblivious to that fact and grants the same reward of 1 for solving every test in T, which may result in premature convergence discussed in Sect. 2.4. In order to entice a search process to pass the more difficult tests, one might want to increase the rewards for them. But where to look for reliable information on test difficulty? The exact objective difficulty (2.2) and subjective difficulty introduced in Sect. 2.2.2 are of little use here: the former requires running all programs in $\mathcal{P}$ on a given test, and the latter estimating the probability that a given synthesis algorithm produces a program that passes a given test.

To estimate the difficulty of particular tests in T, IFS uses the outcomes of their interactions with the candidate programs in the working population $P \subset \mathcal{P}$, and defines the evaluation function as follows:

$$f_{\text{IFS}}(p) = \sum_{t \in T : g(p,t)} \frac{1}{|P(t)|} \qquad (4.4)$$

where $P(t) \subseteq P$ denotes the subset of population members that pass test t:

$$P(t) = \{p \in P : g(p,t)\}. \qquad (4.5)$$

Notice that $P(t)$ corresponds to a column in an interaction matrix (4.3), and $|P(t)|$ is equal to a sum of such a column.

In contrast to evaluation functions considered so far, f_{IFS} is *maximized*. The denominator in Formula 4.4 never becomes zero, because if p passes a given t, then $P(t)$ must contain at least p. The computational overhead of calculating f_{IFS} is usually negligible, because to get evaluated, the programs in P have to be applied to the tests in T anyway.

Example 4.1. Consider a population of three programs $P = \{p_1, p_2, p_3\}$ evaluated on four tests $T = \{t_1, t_2, t_3, t_4\}$, with interaction matrix shown in the left part of Table 4.1. Although p_1 and p_2 pass the same number of tests, p_1 is granted greater value of f_{IFS} because it passes the tests that no other program in P passes. On the other hand, p_2 is not unique in P in its capability of passing t_3. Thus, $f_{\text{IFS}}(p_1) > f_{\text{IFS}}(p_2)$, even though $f_o(p_1) = f_o(p_2)$. ∎

implicit fitness sharing

Table 4.1: Calculation of IFS evaluation for an exemplary population P and four tests in T. The upper left 3×4 part of the table presents the matrix G of interaction outcomes between P and T. The bottom row shows the number of programs in P that pass a given test. The column marked $f_o(p_i)$ presents the conventional objective function, i.e. the number of failed tests. The rightmost column shows the calculation of IFS evaluation, which results from sharing the rewards for solving particular tests. Note that an individual's evaluation is simply the scalar product of its outcome vector with the vector of inverted cardinalities of $P(t)$s.

G	t_1 t_2 t_3 t_4	$f_o(p_i)$	$f_{\text{IFS}}(p_i)$
p_1	1 1 0 0	2	$1+1+0+0 = 2$
p_2	0 0 1 1	2	$0+0+\frac{1}{2}+1 = \frac{3}{2}$
p_3	0 0 1 0	3	$0+0+1+0 = 1$
$|P_{t_i}|$	1 1 2 1		

The key characteristics of IFS is that it estimates difficulty from an evolved population of programs, i.e. a sample that is biased by a specific selection pressure. The term $\frac{1}{|P(t)|}$ in (4.4) is IFS's measure of difficulty of test t, which depends reciprocally, and thus non-linearly, on the number of programs that pass t (contrary to objective test difficulty (2.2)). As a consequence, tests in IFS can be likened to limited resources: individuals in a population *share* the rewards for solving them, where a reward can vary from $\frac{1}{|P|}$ to 1 inclusive. Higher rewards are granted for tests that are rarely passed by population members (small $P(t)$), and lower for the tests passed frequently (large $P(t)$). Allocation of rewards depends on the capabilities of the current population and is in this sense *relative* rather than objective or subjective. Despite this transient nature, empirical evidence shows that f_{IFS} can substantially improve performance compared to the conventional objective function f_o [114, 98].

relative test difficulty

The relative nature of f_{IFS} makes it different from conventional evolutionary algorithms, where an evaluation of a candidate solution is normally *context-free*, i.e. does not depend on the other candidate solutions. IFS may thus seem to resemble a coevolutionary algorithm (Sect. 4.1). However, in coevolutionary algorithms, individuals interact with each other directly, while in IFS there is no face-to-face competition between them. Interestingly, IFS can be also remotely related to *shaping*, an extension of the conventional reinforcement learning paradigm [173]: by varying the rewards for solving particular tests, IFS can be said to modify its own *training experience* [175].

diversity main-tenance
explicit fitness sharing

Because IFS increases the survival odds for candidate solutions that have 'rare competences', it is commonly considered as a diversity maintenance technique and a means of avoiding *premature convergence*. These characteristics motivated also *explicit fitness sharing* proposed in [41], where population diversity is encouraged by monitoring genotypic or phenotypic distances between individuals. By allowing the same program to receive different evaluation in particular generations of an evolutionary run, IFS may also facilitate escaping from local minima.

IFS assumes interaction outcomes to be binary: the tests that have been passed by a program need to be clearly delineated from those that have not. In real-valued domains, that concept is in a sense 'fuzzified' and programs can perform better or worse on individual tests. In [98] we proposed a generalized variant of IFS that ranks programs in a population with respect to errors they commit on a given test and obtain so reliable information on test difficulty. The method achieved top accuracy when confronted with several other extensions of GP on a nontrivial real-world task of detection of blood vessels in retinal imagining.

4.3 Promoting combinations of skills via cosolvability

A program's capability in passing a given test can be likened to a *skill*. We presented that perspective in Sect. 2.3. It reverberates with some earlier works, albeit within very different formal frameworks (for instance production rules, conditional programs, and analogical problem solving in [159]). IFS defines evaluation as a sum of rewards for mastering individual skills. In real world however, it is often the *combination* of skills that matters. Reaching for a biological analogy, the skill of digging in the ground and the skill of navigation may each on its own bring only marginal benefits for an animal. However, when combined, they enable finding previously-buried prey and hence survival when food is scarce, an advantage which can be greater than the sum of the constituent benefits. As another example, the overall performance of a mobile robot may depend on multiple skills, including the ability to maintain a straight-line trajectory, the ability to turn, and the ability of position estimation. Each of these skills alone may be insufficient to complete a given task, but together they may make that possible.

In IFS, the reward for passing two tests simultaneously amounts to the sum of rewards obtained for passing each test individually (4.4). IFS cannot thus model *synergy*, i.e. reward a combination of skills higher than the sum of rewards of its constituents. To model non-additive interactions between skills, in [94] we introduced the notion of *cosolvability*. We call a pair of tests (t_i, t_j) *cosolvable by* a program p if and only if p passes both of them, i.e. $g(p, t_i) \wedge g(p, t_j)$. The *cosolvability matrix* for a population P evaluated on tests in T is a symmetric $|T| \times |T|$ matrix C, with the elements defined as

$$c_{ij} = |\{p \in P : g(p, t_i) \wedge g(p, t_j)\}|, \qquad (4.6)$$

We define then the *cosolvability evaluation function* f_{CS} that rewards programs for solving pairs of *distinct* tests:

skill

synergy of skills

cosolva-bility

cosolvable tests

Table 4.2: An interaction matrix G for an exemplary population of four programs and for four tests (a), and the corresponding cosolvability matrix C (b). Empty cells denote zeroes.

(a)

G	t_1	t_2	t_3	t_4
p_1	1	1	0	0
p_2	0	0	1	1
p_3	0	1	1	0
p_4	1	0	0	1

(b)

C	t_1	t_2	t_3	t_4
t_1	a+d	a		d
t_2		a+c	c	
t_3			b+c	b
t_4				b+d

Table 4.3: Fitness values assigned to programs from Table 4.2a by particular evaluation functions.

Evaluation function	p_1	p_2	p_3	p_4
f_o	2	2	2	2
f_{IFS}	$\frac{1}{a+d} + \frac{1}{a+c}$	$\frac{1}{b+c} + \frac{1}{b+d}$	$\frac{1}{a+c} + \frac{1}{b+c}$	$\frac{1}{a+d} + \frac{1}{b+d}$
f_{cs}	$\frac{1}{a}$	$\frac{1}{b}$	$\frac{1}{c}$	$\frac{1}{d}$

$$f_{\mathrm{cs}}(p) = \sum_{i<j, c_{ij}>0} \frac{1}{c_{ij}} \qquad (4.7)$$

As f_{IFS}, f_c is *maximized*. Similarity of this formula to (4.4) is not incidental: cosolvability can be viewed as a form of second-order fitness sharing, it is the rewards for solving *pairs* of tests that is shared.

Example 4.2. Consider four programs p_1, p_2, p_3, p_4 that perform on tests t_1, t_2, t_3, t_4 as shown in the interaction matrix in Table 4.2a. Assume that the population P contains a programs that produce the same outcome vector as p_1, i.e. a 'behavioral clones' of p_1. Similarly assume b behavioral copies of p_2, c copies of p_3, and d copies of p_4. The cosolvability matrix C for this population is shown in Table 4.2b. Note that co-occurrence of multiple programs that have the same outcome vector is likely in a population of programs that have been evolving for some time.

Table 4.3 presents the evaluations for programs $p_1 \ldots p_4$ as assigned by particular evaluation functions: conventional f_o (1.7), fitness sharing f_{IFS} (4.4), and cosolvability evaluation function f_{cs} (4.7). We note that f_o does not discern programs at all, no matter how often they occur in the population. Whether f_{IFS} and f_{cs} discern particular pairs of programs depends on the numbers of occurrences of $t_1, \ldots, t_4$, i.e. on of a, b, c and d.

Programs p_1 and p_3 allow us to demonstrate that f_{cs} can produce different ordering of individuals than fitness sharing. Let us see if $f_{\mathrm{IFS}}(p_1) < f_{\mathrm{IFS}}(p_3)$ and $f_{\mathrm{cs}}(p_1) > f_{\mathrm{cs}}(p_3)$ can hold simultaneously. As it follows from Table 4.2, these two conditions are respectively equivalent to $a + d > c + b$ and $a < c$, which are fulfilled by infinitely many quadruples of $a, b, c, d \geq 0$. Therefore, f_{CS} can order solutions differently from f_{IFS} and, in consequence, lead to different outcomes of a GP run. ∎

Let us now investigate the ability of f_{IFS} to model synergy between skills. Let $T(p)$ denote the set of tests from T that are passed by p, i.e,

$$T(p) = \{t \in T : g(p, t)\}. \tag{4.8}$$

$T(p)$ corresponds to a row of an interaction matrix (4.3), in analogy to $P(t)$ (4.5) that corresponds to a column.

Consider two programs p, p' such that $T(p) \cap T(p') = \emptyset$. Assume they are crossed over and produce an offspring p_o that is their perfect 'behavioral mixture', i.e. inherits all skills from them and does not have any other skills. Formally,

$$T(p_o) = T(p) \cup T(p'). \tag{4.9}$$

It obviously holds that $f_o(p_o) = f_o(p) + f_o(p')$, because f_o simply counts the passed tests. f_{IFS} is similarly additive and an analogous relation $f_{\mathrm{IFS}}(p_o) = f_{\mathrm{IFS}}(p) + f_{\mathrm{IFS}}(p')$ holds, though for this to be true we need to assume that p_o, p, and p' are members of the same population for which the subjective difficulty of tests is estimated. For the sake of argument, we will stick to this assumption this for the rest of this section.

In contrast to f_o and f_{IFS}, CS is not additive in the above sense. The offspring not only inherits the scores earned by its parents, but also receives additional rewards for passing the pairs of tests the parents have individually failed. In effect, it is guaranteed that $f_{\mathrm{CS}}(p_o) > f_{\mathrm{CS}}(p) + f_{\mathrm{CS}}(p')$. Thus, cosolvability not only enables, but actually *enforces* synergy: an offspring that inherits all skills from parents that have mutually exclusive skills is by definition better than both of them taken together. Given the relative nature of cosolvability, the actual differences in evaluation vary depending on the skills of other programs in a population, nevertheless the above statement is guaranteed to hold.

Now consider two programs p, p' such that $T(p) \neq T(p')$ and $f_{\mathrm{CS}}(p) > f_{\mathrm{CS}}(p')$, and a test t such that $t \notin T(p) \cup T(p')$. Assume that, as a result of modification, both p and p' acquire the skill of passing t, so that for the respective resulting offspring programs o and o' it holds $T(o) = T(p) \cup \{t\}$ and $T(o') = T(p') \cup \{t\}$. From the above analysis it follows that $f_{\mathrm{CS}}(o) < f_{\mathrm{CS}}(o')$ is possible. Thus, o' can gain more than o for passing the same test, to the extent that it becomes better than o, even though its parent was worse than the parent of o. Neither the conventional objective function f_o nor IFS allow for that; under both these evaluation functions, o is better than o'.[1]

---

[1] These observations hold for ordinal selection methods that care only about the ordering of solutions (e.g., tournament selection). For selection methods that

The above properties cause the dynamics of an evolutionary search under f_{CS} to be in general different from that of f_{IFS} and f_o. The differences stem not only from the co-occurrence of skills in a population, but also from the sizes of P and T which determine the likelihood of ties on evaluation. As we noticed in Sect. 2.1, f_o can return only $|T| + 1$ distinct values, so if $|P| \gg |T|$, ties on f_o become likely. For IFS and cosolvability, a complementary relationship holds: the greater the number of programs in P, there more likely it is that different tests are passed by different numbers of programs and, as a consequence, programs are granted different evaluations. In general, ties are thus less likely for IFS than for the conventional evaluation function, and even less likely for cosolvability.

The synergistic nature and fine-grained codomain of f_{CS} proved beneficial in the empirical examination we reported in [94], where, with exception of one out of eight benchmarks, it improved the likelihood of successful program synthesis in comparison to f_o and f_{IFS}. We are aware of only one technical inconvenience of this approach: the size of cosolvability matrix is quadratic with respect to the number of tests, so its memory occupancy may become noticeable when the number of tests reaches the order of thousands.

A concept vaguely related to cosolvability was subject of the study by Lasarczyk et al. [109]. The authors proposed there a method of test selection that maintains a weighted graph that spans tests, where the weight of an edge reflects the historical frequency of a pair of tests being passed simultaneously. The graph is analyzed to select the 'essential' tests that are then used to evaluate all individuals in population. Compared to that approach, cosolvability is a simpler, parameter-free approach, which does not *select* the tests but *weighs* pairs of them, and does that individually for each evaluated program.

4.4 Deriving objectives from program-test interactions

Pareto coevolution

elementary objective

The concept of interaction matrix (4.3) naturally leads to the idea of *Pareto coevolution* [31, 137], where aggregation of interaction outcomes is abandoned in favor of treating each test as an *elementary objective* and comparing candidate solutions with *dominance relation*, as we did in Sect. 2.2.3 with lattices of outcome vectors (Fig. 2.1). A candidate solution p_1 dominates p_2 if and only if it performs at least as good as p_2 on all tests, and strictly better on at least one test. For instance, p_2 in Table 4.1 dominates p_3 as it passes all the tests passed by p_3 and t_4, which p_3 does

---

assume evaluation to be defined on a metric scale (like fitness-proportionate selection), it becomes even easier for f_{CS} to produce evaluations that imply different selection probabilities than those of f_{IFS}.

not pass. On the other hand, there is no dominance between p_1 and p_2 – none of these programs is clearly better than the other.

In principle, dominance relation on tests (*dominance on tests* in the following) can be directly used to determine the outcomes of selection in an evolutionary loop of GP [88]. The arguably simplest selection operator of this kind would, given a pair of programs, return the one that dominates the other, or pick any of them at random in case of mutual non-dominance. However, when the number of tests is large, dominance between candidate solutions becomes unlikely, as there is high chance that each of compared solutions passes a test that the other solution fails. The dominance relation becomes sparse, with many pairs of candidate solutions left incomparable. This in turn weakens the search gradient, and makes search process less effective.

The limitations of dominance on tests as a means for selection of candidate solutions sparked search for alternative means of exploiting interaction matrices. The breakthrough came with the observation that test-based problems may feature an *internal structure*. Bucci [16] and de Jong [23] introduced *coordinate systems* that *compress* the elementary objectives (each associated with a unique test) into a multidimensional structure of *underlying objectives* (dimensions), while preserving the dominance relation between candidate solutions. Because some tests can be redundant, the number of underlying objectives can be lower (and, interestingly, may indicate the inherent complexity of a given test-based problem).

However, coordinate systems do not address the above problem of dominance on tests being sparse (or becoming sparse in the course of search). As a coordinate system perfectly preserves dominance, whenever the dominance on tests is sparse, so it is in the dominance on the underlying objectives derived from them. Also, the number of underlying objectives can be still high, even for simple problems like the game of tic-tac-toe, and construction of a coordinate system is an NP-hard problem [64, 63].

These observations call for alternative ways of efficiently translating an interaction matrix into a computationally tractable multi-aspect characterization of candidate solutions. In [112, 95] we came up with the idea of discovering *approximate* objectives by heuristic clustering of interaction outcomes. The proposed method DOC efficiently clusters an interaction matrix into a low number of performance measures, which we refer to as *derived objectives*, to clearly delineate them from the exact underlying objectives. By corresponding to a subset of tests, each derived objective captures a 'capability' that can be seen as a generalization of skills discussed earlier.

Technically, DOC replaces the conventional evaluation stage of the GP workflow (cf. Sect. 1.5.3) with the following steps:

[margin notes:] dominance relation on tests — coordinate systems — underlying objective — derived objective

1. *Calculation of interaction matrix.* We apply every program in the current population P, $|P| = m$, to every tests in T, $|T| = n$, and obtain so an $m \times n$ interaction matrix G (4.3).

2. *Clustering of tests.* We treat every column of G, i.e. the vector of interaction outcomes of all programs from P with a test t, as a point in an m-dimensional space. A clustering algorithm of choice is applied to the n points obtained in this way, and produces a partition $\{T_1, \ldots, T_k\}$ of the original n tests in T into k subsets (clusters), where $1 \leq k \leq n$ and $T_j \neq \emptyset$.

3. *Calculation of derived objectives.* For each cluster T_j, we average rowwise the corresponding columns in G. The result is an $m \times k$ *derived interaction matrix* G', with the elements defined as follows:

$$g'_{i,j} = \frac{1}{|T_j|} \sum_{t \in T_j} g(p_i, t) \tag{4.10}$$

where p_i is the program corresponding to the ith row of G, and $j = 1, \ldots, k$.

The columns of resulting G' matrix define the k derived objectives that characterize the programs in P in the context of the tests in T. The jth derived objective for a program p_i corresponding to the i row of the derived interaction matrix G' amounts to

$$f^j_{\text{DOC}}(p_i) = g'_{i,j}. \tag{4.11}$$

Example 4.3. Figure 4.1 presents the example of DOC deriving objectives from a 4×5 interaction matrix G. The clustering algorithm partitions the tests into $k = 2$ clusters $\{t_1, t_2\}$ and $\{t_3, t_4, t_5\}$. Averaging the corresponding columns in G leads to the 4×2 derived interaction matrix G'. The graph plots the programs' positions in the resulting two-dimensional space of derived objectives. ∎

The derived objectives constructed by DOC form a compact, multi-aspect evaluation of the candidate solutions in P, and serve as a basis for selecting the most promising programs. Rather than devising an ad-hoc selection algorithm, it is natural to employ here multiobjective methods like NSGA-II [26]. Multiobjective selection allows programs that feature different behaviors (capabilities) coexist in a population, even if some of them are better than others on the conventional objective function f_o. In DOC, a capability can be identified with passing a specific group, or even *class* of tests. In case of the parity-3 problem illustrated in Sect. 2.3, a capability could be

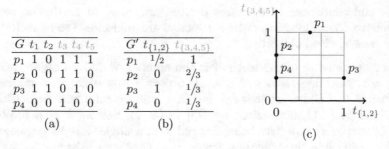

G	t_1	t_2	t_3	t_4	t_5
p_1	1	0	1	1	1
p_2	0	0	1	1	0
p_3	1	1	0	1	0
p_4	0	0	1	0	0

(a)

G'	$t_{\{1,2\}}$	$t_{\{3,4,5\}}$
p_1	$1/2$	1
p_2	0	$2/3$
p_3	1	$1/3$
p_4	0	$1/3$

(b)

(c)

Fig. 4.1: An example of derivation of two search objectives from a matrix G of interactions between four programs $p_1, \ldots, p_4$ and five tests $t_1, \ldots, t_5$ (a). The tests (corresponding to the columns of G) are clustered into two clusters, marked in colors, according to a distance metric (here: Euclidean distance). The centroids of the clusters form the derived interaction matrix G' (b), in which each column defines a derived objective. The derived objectives form new objective space, with programs' locations shown in inset (c).

associated with passing all tests with the first input variable set to *true*. See Sect. 9.8 for a more detailed description of NSGA-II.

Derived objectives bear certain similarity to the underlying objectives discussed at the beginning of this section [16, 23, 64, 63]. However, as Example 4.3 and Fig. 4.1 show, they are not guaranteed to preserve dominance: new dominance relationships may emerge in the space of resulting derived objectives. For instance, given the interaction matrix as in Fig. 4.1a, program p_3 does not dominate p_4, however it does so in the space of derived objectives (Fig. 4.1c). As a result of clustering, some information about the dominance structure has been lost. This inconsistency buys us however a critical advantage: the resulting dominance relation is more dense and thus likely to impose a reasonably strong search gradient on an evolving population.

Although DOC may lead to dominance in the space of derived objectives where such relation was originally absent, in another work under review [113] we show formally that derivation of objectives will always preserve dominance if it already held for a pair of candidate solutions. Also, it cannot reverse the direction of dominance that already existed in the original space of outcome vectors.

Because clustering *partitions* the set of tests T (rather than only *selecting* some of them), none of the original tests is ignored in the evaluation process. In this sense, DOC tends to embrace the entirety of information available in an interaction matrix, which makes it different from and potentially more robust than methods that select tests, like [109] reviewed briefly at the end of Sect. 4.3. The more two tests are similar in terms of programs' performance on them, the more likely they are to end up in the same cluster and contribute to the same derived objective. In particular, tests characterized with identical outcome vectors are guaranteed to be included in the same derived objective.

The only parameter of the method is the number of derived objectives k. For $k = 1$, DOC degenerates to a single-objective approach: all tests form one cluster, and G' has a single column that contains solutions' evaluations as defined by (1.7), normalized by $|T_j|$ in (4.10). Setting $k = n$ implies $G' = G$, and every objective being derived from a single test. As we showed in [112], using k in the order of a few is most beneficial. Alternatively, the choice of k can be delegated to the clustering algorithm [95].

Similarly to IFS and CS, evaluation performed by DOC is contextual: all programs in P together determine the values of derived objectives. Objectives are derived independently in every generation of a GP run and are thus transient and incomparable across generations. This however does not prevent them from driving search more efficiently than conventional GP and IFS on most benchmarks, which we demonstrated in [95], and in coevolutionary settings, where T varies from generation to generation [112].

4.5 Summary

In the context of behavioral evaluation and execution record (Chap. 3), IFS, CS and DOC all rely on the same source of information for evaluation: an outcome vector resulting from the comparison of program output with the desired output (Fig. 3.2). In contrast to the conventional objective function f_o (1.7) that simply counts the zeroes (failed tests) in that vector for the program which is being evaluated, these methods require simultaneous access to outcome vectors of all programs in a population. Only then can they assess the subjective difficulty of tests (IFS), estimate the subjective odds for pairs of tests being simultaneously passed (CS), or group the tests into meaningful clusters to form derived objectives (DOC). In consequence, they will in general lead to different selection outcomes (see Examples 4.1, 4.2, and 4.3).

There is however more information available in an execution record and in the tests that define a program synthesis task. In particular, IFS, CS and DOC care only whether a test has been passed or not, and ignore *what* is the actual program output and the desired program output. These more detailed data open the door to more 'inquisitive' extensions of GP, with semantic GP presented in the next chapter being an important contemporary representative.

5

Semantic Genetic Programming

Semantic genetic programming (SGP) is a relatively new thread in GP research, which originated in the immense complexity of the genotype-phenotype mapping in evolutionary program synthesis. As discussed in Sect. 1.4, minor modifications of program code may result in fundamentally different behavior; on the other hand, an overhaul of program may leave its behavior intact. The relationship between program source code (syntax) and its behavior (semantics for the sake of this chapter) is very complex. SGP germinated from the increasing belief that to make evolutionary program synthesis scalable, program synthesis algorithms need to explicitly take program semantics into account. In this chapter, we provide a concise insight into SGP and show how its conceptual underpinnings relate to behavioral program synthesis and execution records.

5.1 Program semantics

Program semantics is an important concept in theory of programming languages and in design of compilers and other tools used in contemporary software engineering. Formalisms of denotational, axiomatic, and operational semantics have longstanding position in computer science. The notion of semantics adopted in GP is however different in being tailored to test-based framework of program synthesis.

Following the consensus emerging from earlier works on SGP [184, 99, 183, 35, 145], we define program semantics as follows. Given a list of tests T, $|T| = n$, the *semantics of a program* p is an (ordered) n-tuple[1] of the outputs produced by p for the tests in T, i.e.

program
seman-
tics

---

[1] Even though program semantics is a *vector* for numeric domains, in the following we refer to them as tuples to emphasize that programs can return data of any type. Tuples are assumed to be ordered.

© Springer International Publishing Switzerland 2016
K. Krawiec, *Behavioral Program Synthesis with Genetic Programming*,
Studies in Computational Intelligence 618,
DOI: 10.1007/978-3-319-27565-9_5

$$s_T(p) = (p(in))_{(in,out)\in T} = (p(in_1), p(in_2), \ldots, p(in_n))\tag{5.1}$$

Program semantics is thus a part of an execution record (3.5) and can be alternatively defined by referring to the final execution state $p^\$(in)$ (Fig. 3.2):

$$s_T(p) = (p^\$(in))_{(in,out)\in T}.\tag{5.2}$$

Note that (5.2) holds in the strict sense only if the *entire* final execution state can be interpreted as program output, which is for instance true in side effect-free tree-based GP. For other program representations like PushGP [170] or linear GP [6, 14], program output (and thus $s_T(p)$) comprises only selected elements of the final execution state (e.g., top stack elements in PushGP or dedicated registers in linear GP).

In practice, semantics is most often based on the tests T given with a synthesis task, which in the following we assume to be known implicitly and thus write $s(p)$ instead of $s_T(p)$.

This rephrasing in terms of execution records helps convey that $s(p)$ reflects only the final execution states. It is partial also in another sense: as $s(p)$ involves only the tests available in T (which is usually only a sample of a universe of tests $\mathcal{T}$), it is agnostic about program behavior beyond that set. This makes it different from more formal operational, axiomatic or denotational semantics. In an attempt to emphasize that fact, the term 'sampling semantics' [136] has been alternatively promoted for (5.1); nevertheless, for brevity we call this formal object simply 'semantics'.

semantics Program semantics characterizes an existing program. For the sake of further argument, a broader notion of semantics will be necessary. Given a set of tests $T \subseteq \mathcal{I} \times \mathcal{O}, n = |T|$, we define *semantics s* as an *arbitrary n*-tuple of elements of $\mathcal{O}$, i.e. $s \in \mathcal{O}^n$. As in (5.1), an element s_i of s corresponds to the ith test in T, but this time in abstraction from any program. A program producing s_i for the ith test does not have to exist in a given programming language. In this way, semantics may express any combination of behaviors on all tests in T. In particular, the target t^* (1.4), i.e. the tuple of desired outputs as defined by a program synthesis task, is also a semantics.

semantic space semantic mapping The set of all semantics for a set of tests $T \subseteq \mathcal{I} \times \mathcal{O}$ forms the *semantic space* $\mathcal{S}$. In this context, s can be seen as a function, a *semantic mapping* $s : \mathcal{P} \to \mathcal{S}$. Every program semantics is a semantics, because by definition $s(p) \in \mathcal{S}$ for any p. The reverse is not true though: some semantics in $\mathcal{S}$ may form such combinations of elements from $\mathcal{O}$ that cannot be produced by any program in $\mathcal{P}$. Semantic mapping s is thus a non-surjective function from $\mathcal{P}$ to $\mathcal{S}$.

Example 5.1. Figure 5.1a reprints the symbolic regression program p from Fig. 3.1. Assume a set of two tests T as in Fig. 5.1b. The domain of the programing task specified by T is $(\mathcal{I}, \mathcal{O}) = (\mathbb{R}^2, \mathbb{R})$. Given two tests with

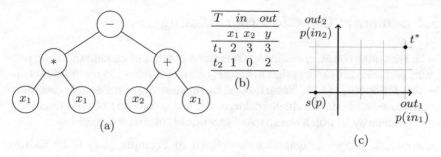

(a)

T	in		out
	x_1	x_2	y
t_1	2	3	3
t_2	1	0	2

(b)

(c)

Fig. 5.1: An exemplary program p reproduced from Fig. 3.1 (a) and a list of two tests T, each comprising two input variables x_1 and x_2 (b). The program outputs respectively -1 and 0 for these tests, so its semantics is $s(p) = (-1, 0)$. The inset (c) shows the semantic space $\mathcal{S}$ associated with T and the locations of $s(p)$ and T's target t^* in that space.

outputs in $\mathbb{R}$, the semantic space is $\mathcal{S} = \mathbb{R}^2$. Figure 5.1c presents the semantic space with the marked semantics of program p, $s(p) = (-1, 0)$, determined by applying p to the inputs of tests in T. The target $t^* = (3, 2)$ as defined by T is also marked in that space.

Assume the set of instructions of the programming language is $\{+, *, /\}$. The target $t^* = (3, 2)$ is realizable in this programming language, i.e. a program p^* exists such that $\forall(in, out) \in T : p^*(in) = out$. As signaled earlier, in general a program realizing a given semantics $s \in \mathcal{S}$ does not have to exist. Let us illustrate two cases of such *unrealizable semantics*:

unrealizable semantics

Case 1: Under the same programming language, $s' = (-1, 2)$ is unrealizable because no program composed of instructions $\{+, *, /\}$, no matter how complex, can produce a negative value for the positive arguments $x_1 = 2, x_2 = 3$ in the first test t_1.

Case 2: Let the programming language be $\{+, *\}$. Then, $s'' = (2, 3)$ is unrealizable, because the programs composed of these instructions can express only monotonically increasing functions. As x_1 and x_2 are both greater in t_1 than in t_2, any program composed of these instructions ensures $p(in_1) > p(in_2)$. ∎

The example illustrates that whether a program with a given semantics exists or not depends on the expressibility of programming language and the constraints imposed by tests. Inexpressibility of a given semantics may stem from certain elements in $\mathcal{O}$ being impossible to generate using the available instructions (Case 1), or the particular *combination* of outputs being impossible to achieve given the inputs in T (Case 2). Nevertheless, even if a semantics is realizable, it can be challenging to synthesize a program that realizes it; we address this problem in the next section.

5.2 Semantic Genetic Programming

In the previous chapter, evaluation functions in IFS and CS characterized program performance in ways that are clearly different from the conventional objective function f_o (1.7). Nevertheless, both these approaches still compress program behavior into a single scalar, and in this sense do not fundamentally change the way in which the space of candidate solutions is searched.

In contrast, program semantics as defined in Formula (5.1) is by nature multi-dimensional: each element of semantics $s(p)$ characterizes p's behavior for a given test. This detailed behavioral characterization opens the door to defining semantic-aware components of GP workflow, including semantic-aware population initialization, selection, and search operators, with the last area studied most intensely. As semantic GP is one among several takes on behavioral program synthesis presented in this book, we review here only the most relevant contributions and refer the reader to more comprehensive surveys in [145, 144, 142].

Apart from analytical works conducted by, among others, Langdon [104], McPhee et al. were to our knowledge the first to study the impact of crossover on program semantics [125]. In that work, they conceptualized semantic building blocks, defined the semantic properties of components that form offspring in the tree-swapping crossover, i.e. subtrees and contexts (partial trees with a hanging branch), and observed how they change with evolution for Boolean problems. In another early study, Beadle and Johnson [10] proposed a semantically driven crossover operator for Boolean problems that guarantees the offspring to be semantically distinct from both parents. In [62] Jackson analyzed semantic diversity in the initial population. More recently, Galvan-Lopez et al. came up with a tournament selection that discouraged semantic duplicates [35].

Most of the work on semantic-aware search operators revolved around the observation that semantic similarity of programs (as well as subprograms) can be measured by referring to the corresponding points in semantic space. Nguyen et al. [184] considered two semantic crossover operators for symbolic regression, one that permits crossover only if the subtrees to be exchanged in the parents are semantically more distant than a given lower limit (a parameter of the method), and another with an additional upper limit on the distance. Later, Nguyen et al. [183] proposed an operator that, from a set of all valid pairs of subtrees to be exchanged in the parents, chooses the pair with the smallest distance, dropping so the upper limit parameter. Driven by similar intentions, though without explicitly referring to program semantics, Day and Nandi used binary strings to characterize how individuals in a population cope with particular tests, and designed a mating strategy that exploits that information [22]. More recently, we proposed semantic backpropagation [145] that enables partial reversal of program execution and so facilitates design of effective search operators.

5.3 Geometric Semantic Genetic Programming

The work reviewed above was driven by the legitimate assumption that per-test information on program behavior may help in driving search more efficiently by, among others, avoiding the compensation discussed in Sect. 2.2.2. The experimental outcomes were in most cases encouraging. Nevertheless, pure SGP brought no major conceptual breakthroughs.

The new era in SGP came with the realization that there are deeper implications of posing a program synthesis task as a search for a program with a certain semantics, rather than for a program with a certain evaluation. As it turns out, the test-based objective functions typically used in GP (e.g., MSE, MAD; see Sect. 1.5.3) are formally *metrics* in $\mathcal{S}$. By this token, they formally turn the set $\mathcal{S}$ into a *space* with certain geometry that can be exploited for the sake of search. This observation gave rise to *geometric semantic* GP (GSGP) that focuses specifically on the geometric (metric-related) properties of semantics.

geometric
semantic
GP

Formally, *semantic metric* is any function

semantic
metric

$$d : \mathcal{S} \times \mathcal{S} \to \mathbb{R}^{0+} \tag{5.3}$$

that is non-negative, symmetric, and fulfills the properties of identity of indiscernibles and triangle inequality. In order to demonstrate that the evaluation functions typically used in GP are indeed semantic metrics, consider the Boolean domain and assume that d is the Hamming distance d_H. Then, the objective function f_o (1.7) for a program synthesis task with a target t^* can be rewritten as

$$f_o(p) = d_H(s(p), t^*). \tag{5.4}$$

With $\mathcal{S}$ being a metric space, the evaluation at any point in that space is the distance from the target t^*, i.e. from the desired semantics as specified by the synthesis task. Thus, the surface of an evaluation function plotted with respect to $\mathcal{S}$ takes the form of a cone with the apex corresponding to t^*. The specific type of the cone depends on the employed metric; for d being the Euclidean distance, it is a cone in the common sense of that word.

Crucially, f_o is not only conic but also unimodal and attains the minimum value of zero at t^* and nowhere else. It also does not feature plateaus. This holds for any semantic space, any data type associated with the output domain $\mathcal{O}$, and any metric, with the caveat that the cone may have an unintuitive interpretation in the non-Euclidean spaces. A visualization of an evaluation function spanning $\mathcal{S}$ under the city-block metric and the Euclidean metric is presented in grayscale in Fig. 5.2.

Although minimizing f_o in such a simply structured space may appear easy, it is not such in practice, the reason being that the semantic space $\mathcal{S}$ *is not the space being searched.* The search space is the set of programs $\mathcal{P}$, because it is programs, not semantics, that are manipulated by search operators. $\mathcal{S}$ is searched only implicitly, intermediated by the semantic mapping $s : \mathcal{P} \to \mathcal{S}$. As a result, applying a minor change to a program can correspond to a big leap in $\mathcal{S}$, while a major change of a program can leave its semantics intact. In other words, the cone in question is *not* a fitness landscape in the traditional meaning of that term (Sect. 2.2.3, [194]), because the moves in $\mathcal{P}$ induced by search operators do not correspond to the moves along the dimensions of $\mathcal{S}$. This characteristic is analogous to low *locality* of genotype-phenotype mapping discussed in Sect. 1.4.

Despite detachment from fitness landscapes, semantic space gives certain hints for designing search operators, and GSGP offers a formal framework for principled design of efficient search operators under given semantic metric d. GSGP determines the desired semantic properties of the offspring produced by a search operator *in abstraction from any specific search operator*. Our recent formalization of such properties in [135], together with earlier attempts [93], delineate the class of *geometric semantic search operators*, including geometric semantic mutation and geometric semantic crossover.

geometric
semantic
search
opera-
tors

Here, we limit our discourse to geometric semantic crossover, because prospectively it offers the greatest leverage, enabling long-distance moves in the semantic space and substantially improving convergence to global optima. In a more general context, let us state that the emphasis on crossover is consistent with our general stance, i.e. that, as argued elsewhere [2], recombination is the key innovation of evolutionary computation. Without it, an evolutionary algorithm 'degenerates' to a parallel local search and has limited chance for benefiting from *modularity*, which is often present in the structure of the task and search space (see more in-depth discussion on modularity in Sect. 11.1). In this sense, we find it appropriate to 'rehabilitate' crossover, which since the Schema Theorem has been all too often blamed for its 'disruptive' character, though in fact its overall role may be much more constructive than widely assumed –see for instance the line of argument concerning Royal Road benchmarks in Sect. 4.2 of [132].

The core innovation of GSGP comes with the observation that a semantic metric d can be used not only to measure the distance of program's semantic from the target (5.4), but to compare *any* semantics. An offspring program $p = c(p_1, p_2)$ resulting from an application of a crossover operator $c : \mathcal{P} \times \mathcal{P} \to \mathcal{P}$ to a pair of parent programs (p_1, p_2) is *geometric with respect to p_1 and p_2* under metric d if and only if its semantics is located in the d-metric segment connecting the semantics of its parents, i.e.

geometric
offspring

$$d(s(p_1), s(p_2)) = d(s(p_1), s(p)) + d(s(p), s(p_2)). \tag{5.5}$$

A crossover operator that guarantees (5.5) for any pair of parents and resulting offspring is called *geometric semantic crossover*, or *geometric crossover*

geometric
crossover

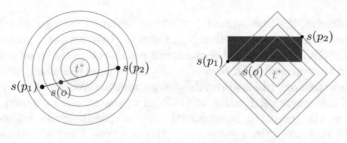

Fig. 5.2: Illustration of semantic space (in grayscale) and geometric crossover (in color), for the Euclidean metric (left) and for the city-block metric (right), in two-dimensional semantic space (i.e. for two tests). Spatial dimensions (abscissa and ordinate) correspond to outputs for the first and the second test, respectively. The target t^* determines the desired semantics as posed by a program synthesis task. For crossover, $s(p_1)$ and $s(p_2)$ mark the semantics of parent programs p_1 and p_2; $s(o)$ marks the semantics of their exemplary geometric offspring o. The colored segments spanning $s(p_1)$ and $s(p_2)$ define the set of semantics of offspring that are geometric with respect to the parents.

for short. This definition formalizes the characteristic that is considered desirable in crossover, namely that an offspring should have some 'traits' in common with both its parents. In semantic GP, a trait of a program p can be identified with the output it produces for a test, i.e. with an elements of its semantics $s(p)$. Traits are in this context tightly related to skills (Sects. 2.4 and 4.3), the difference being that a skill is a binary indicator that communicates only the passing or failing of a test, while a trait in the context is a specific *output* produced by a program.

The concept of geometric offspring for city-block metric and the Euclidean metric is presented in color in Fig. 5.3. The segments mark the locations of potential offspring that are geometric with respect to the parents p_1 and p_2. The point marked $s(o)$ is the semantics of a hypothetical offspring o. Clearly, geometric offspring have high probability of getting closer to the target t^* and thus obtaining better evaluation, particularly when the parents happen to lie on opposite slopes of the cone of evaluation function. In particular, for the Euclidean distance, the geometric offspring is guaranteed to be at least as good as the worst of the parents. For this and other types of guarantees that may apply to geometric offspring, see [143].

By belonging to the segment (5.5), a geometric offspring is guaranteed to minimize the total distance (dissimilarity) to its parents in semantic space. Note that this characteristics is unrelated to *equidistance* from the parents, i.e.

$$d(s(p_1), s(p)) = d(s(p), s(p_2)), \tag{5.6}$$

which has been studied in the past in EC [126] and GP [86].

To our knowledge, the first semantic crossover designed with geometric aspects in mind was KLX proposed in [93]. Later in [135], an exact geometric crossover (GSGX) was proposed that guarantees the offspring to be strictly geometric in the sense of (5.5). Recently, [99] introduced the locally geometric semantic crossover, which approximates geometric crossover 'locally', on the level of subprograms. Finally, in [145] an operator was proposed that approximates the geometric recombination by propagating the desired semantics of an offspring through parent's program tree. Of these operators, we present here in detail KLX and GSGX. For a full review of state-of-the-art geometric semantic search operators, the reader is referred to [144, 142].

5.3.1 Approximate geometric crossover

approximate geometric crossover

The KLX crossover proposed in [93] uses an arbitrary *base crossover operator* to repetitively produce candidate offspring from the same parents, and by this token can be considered a form of *brood selection* [177]. Given two parent programs p_1 and p_2, KLX applies the base crossover operator k times to them and stores the resulting candidate offspring in a *breeding pool*. Then, it calculates the following expression for every candidate offspring p:

$$d(s(p_1), s(p)) + d(s(p), s(p_2)) + |d(s(p_1), s(p)) - d(s(p), s(p_2))|. \quad (5.7)$$

The two candidate offspring in the breeding pool that have the lowest value of (5.7) are returned as the final outcome of crossover act. The base crossover operator has to be obviously stochastic, otherwise this trial-and-error approach would not make sense. In [93], the conventional tree swapping crossover [79] served that purpose.

The two first terms in Formula (5.7) are candidate offspring's distances from the parents. The sum of these terms captures the 'degree of geometricity' of the offspring p. The lower this sum, the closer is p to the d-segment connecting p_1 and p_2. Geometricity achieves the minimum for the strictly geometric offspring, i.e. for p located on that segment (cf. (5.5)), including the ends of that segment, i.e. $s(p_1)$ and $s(p_2)$. Because producing candidate offspring that are semantically equivalent to one of the parents renders crossover ineffective in semantic terms, 5.7 involves also the third term that promotes the candidate offspring that is close to being semantically equidistant (c.f. 5.6). An ideal offspring according to (5.7) is thus a program that is both geometric with respect to its parents and equidistant from them. The reader is referred to [93] for a more detailed description of KLX (called SX+ in that paper).

In relying on stochastic trial-and-error, KLX is not guaranteed to produce a geometric offspring. However, it is arguably an *approximate* geometric semantic crossover, because the odds for returning an exactly geometric offspring grow with the size of the breeding pool. Unless the base crossover operator is principally unable to produce a geometric offspring or the particular pair of parents has no geometric offspring at all, a geometric offspring will eventually be generated. Generating large pools of candidate offspring programs is however computationally prohibitive, as each of those programs needs to be run on all tests to calculate (5.7).

5.3.2 Exact geometric crossover

The exact geometric semantic crossover operator (GSGX, [135]) does not involve trial-and-error generation of candidate offspring and so avoids an excessive computational cost. Given two parent programs p_1, p_2, GSGX randomly generates a random program p_r, and then combines p_1, p_2 and p_r into an offspring p using the following formula for the Boolean domain

$$p = (p_1 \wedge p_r) \vee (p_2 \wedge \overline{p_r}),$$ (5.8)

and the following one for the regression domain:

$$p = p_1 * p_r + p_2 * (1 - p_r).$$ (5.9)

The operators are illustrated in Fig. 5.3. The key feature of GSGX lies in recombining the parents using language constructs composed of instructions from the programming language of consideration. To emphasize this, in the Fig. 5.3 we clearly delineate the parent programs (triangles) and the instructions used to combine the parents (ovals). The instructions in question need to be available in the programming language; for the Boolean domain, these are $\wedge$ and $\vee$, while for the symbolic regression domain $+$, $-$, and $*$.

Analysis of Eqs. 5.8 and 5.9 reveals that the offspring are indeed 'semantic mixtures' of the parents. The offspring generated by GSGX is geometric *by construction* and in this sense *exact* (see [135] for formal proofs). Interestingly, (5.8) and (5.9) do not refer to the semantic mapping s at all (in contrast to, e.g., KLX (5.7). Thus, GSGX does not even need to 'know' parents' semantics to guarantee the offspring to be semantically geometric with respect to them. This does not however mean that GSGX ignores the behavioral aspects of program synthesis; to the contrary, it actually relies on it with its specific, provably geometric, approach to offspring construction that relies on the properties of the underlying semantic space.

For the Boolean domain, the 'mixing' program p_r can be arbitrary. For the regression domain, the offspring is a linear combination of the parents.

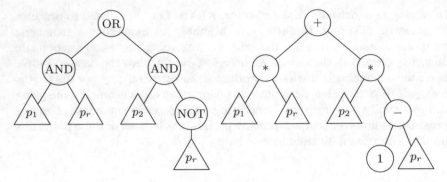

Fig. 5.3: The exact geometric semantic crossover (GSGX) for tree-GP, for the Boolean domain (left) and the symbolic regression domain (right). p_1 and p_2 are parent programs, p_r is a randomly generated subprogram. Ovals mark single instructions.

To make sure that its semantics is located on the segment connecting the parents, p_r must return a value in $[0, 1]$ for every test. If d is the Euclidean distance, p_r has to return the same value in $[0, 1]$ for all tests; for city-block distance, the value returned by p_r should vary across the tests. In [135], we demonstrate how to design a geometric crossover for another domain of rule-based classifiers.

The guarantee of producing an exactly geometric offspring allows GSGX to realize search as a traversal of the above-mentioned cone in semantic space. The consequence is fast convergence to the global optimum and very good performance on conventional GP benchmarks and synthetic problems. This is particularly true when GSGX operates along with an analogously designed *geometric semantic mutation* operator GSGM which guarantees producing offspring that only slightly diverges from its parent in the semantic space. Such mutations obey the structure of semantic space and are essential to precisely hit the target semantics [135].

geometric
semantic
mutation

GSGX and GSGM might be thus thought to be the ultimate tools for semantic GP. However, in each application, GSGX produces offspring that are on average roughly two times larger than its parents. The average number of nodes in a program tree is given by the formula:

$$\overline{|p_0|} + (2^n - 1)(\overline{|p_0|} + \overline{|p_r|}), \tag{5.10}$$

where $\overline{|p_0|}$ is the average number of nodes in a program in the initial population, $\overline{|p_r|}$ is the average number of nodes in the random program p_r (Eqs. 5.8 and 5.9), and n is the generation number. Thus, programs in a search process using GSGX grow exponentially with time. For GSGM, the growth is linear but in a longer run also severe.

To work around this problem, in [135] we proposed to apply simplification to the offspring of every crossover act. Efficient simplification procedures exist for program representations employed there, i.e. disjunctive normal forms for the Boolean domain and vectors of polynomial coefficients for the symbolic regression domain. For the tree-based programs, simplification is known to be NP-hard, in which case heuristic simplifiers like Espresso [155] can be employed.

Syntactic simplification incurs additional computational cost. To mitigate this problem and tackle growing program length in GSGP, Castelli et al. [20] proposed to transform a population of programs into a directed acyclic graph. Because the parent programs get incorporated into the offspring without being modified (Fig. 5.3), GSGX can be applied multiple times to the same parents, and over the course of multiple generations, without ever copying the parents' code. Technically, it is enough for the offspring to *refer* to parents' code rather than to copy it, because they are considered *immutable*. The only new code added in every generation are the expressions that implement the Eqs. (5.8) and (5.9) and the subprograms p_r. As a result, memory consumption grows linearly with the number of generations in this 'implicit memoization' technique.

The characteristics of GSGP discussed here can be linked to *problem decomposition*. GSGP operators tackle a problem in a test-by-test manner, with the mixing trees (p_r's) working as 'filters' that decide about which program subtree's output should be let through to reach offspring's output and which not (in the Boolean case) or combining such outputs linearly (in the continuous case). The test-wise characteristics applies in particular to the SGM mutation operator not discussed here in detail, which produces an offspring that is guaranteed to vary semantically from the parent on exactly one test. This topic will be subject to a broader discussion in Sect. 11.1.

5.4 Summary

The methods of semantic GP form an exception in this book, as they do not explicitly define new, alternative evaluation functions. Rather than that, they rely on the conventional scalar objective function that we have so much complained about in Sect. 2.1. Given that, do they belong to behavioral program synthesis we outlined in Chap. 3?

Our answer to this question is positive. Most of the methods discussed in this chapter and reviewed in cited work require access to semantics of programs (parents, candidate offspring) and semantics of task specification (target, i.e. desired output). Program semantics as meant in GP is a detailed account on program behavior, much richer than scalar evaluation. The only exception are the exact geometric semantic operators, where the guarantees about offspring behavior stem from the geometry of semantic space induced by metric evaluation functions.

By scrutinizing the detailed final effects of computation, semantic-aware methods make search algorithms better-informed about the per-test effects of computation, mitigate so the evaluation bottleneck problem (Chap. 2), and can be thus deemed behavioral. However, final output is arguably a very limited account of behavior of a program, which could have conducted very complex computation prior to arriving at that result. This becomes particularly evident when defining program semantics in terms of execution record (Fig. 3.2): the potentially interesting internal dynamics of program execution is not directly reflected in program semantics. In the approaches presented in the next two chapters, we push the behavioral envelope further, in the limit making the entire evaluation record available to a search process.

6

Synthesizing programs with consistent execution traces

The main motif of this book is providing search algorithms with rich information on solutions' characteristics. The formalism of execution record, a complete, instruction-by-instruction account on program execution for every test, is a technical means to achieve that goal. In the approaches presented to this point, only the final execution states in an execution record were taken into account. In this chapter, we utilize an entire execution record for the first time in this book.

In particular, we illustrate how the concepts borrowed from information theory facilitate deriving alternative evaluation functions from an execution record. To this end, we elaborate on the approach we proposed in [100].

6.1 Information content of execution states

Execution of an instruction of a program usually changes the state of the execution environment (registers, memory, etc.), which then become reflected in an execution record, as we discussed in Chap. 3. At any given stage of program execution, the state of execution environment is characterized by certain *information content*. Crucially, as we show in this section, a deterministic sequential program can at most sustain the amount of information in an execution environment, but is unable to increase it. In consequence, the process of program execution is usually accompanied by gradual loss of information in the environment. This is the key motivation for the evaluation function presented in this chapter.

Example 6.1. As an illustration, consider linear programs running in an execution environment with memory comprising two one-bit registers r_1 and r_2. A program run in this environment has in general the signature

© Springer International Publishing Switzerland 2016
K. Krawiec, *Behavioral Program Synthesis with Genetic Programming*,
Studies in Computational Intelligence 618,
DOI: 10.1007/978-3-319-27565-9_6

$$p = (r_1 \leftarrow AND(r_1, r_2);$$
$$r_2 \leftarrow AND(r_1, r_2))$$

	p^0	p^1	$p^\$$
t_1	00	00	**00**
t_2	01	01	**00**
t_3	10	00	**00**
t_4	11	11	**11**

Fig. 6.1: A linear GP program to be run in an execution environment with two Boolean registers (r_1, r_2) and its execution record for tests that enumerate all inputs. The inputs determine the initial content of the registers, shown in column p^0 (the state before executing the first instruction). Identical execution states are marked in the same color. Bold font in the last column marks the effective output.

$\mathbb{B} \times \mathbb{B} \to \mathbb{B} \times \mathbb{B}$, i.e. $(\mathcal{I}, \mathcal{O}) = (\mathbb{B}, \mathbb{B})$. We assume however that r_1 is designated as program output, so the signature is effectively $p : \mathbb{B} \times \mathbb{B} \to \mathbb{B}$.

Consider an exemplary program runnable in this environment, shown in Fig. 6.1a. The program comprises two instructions: first of them modifies r_1, the second changes r_2. Figure 6.1b presents the execution record of that program for tests enumerating all combinations of inputs (x_1, x_2). With two registers, the state of the execution environment is a tuple of bits (Booleans), which for clarity we present without delimiters, e.g., 01 denotes $(r_1, r_2) = (0, 1)$. Prior to program execution, x_1 and x_2 determine the initial content of registers in p^0. The states in the columns marked p^1 and $p^\$$ reflect the registers after executing the first and the second instruction, respectively. As in the previous examples, there is one-to-one correspondence between the instructions of the program and the columns of execution record $p^1, \ldots, p^\$$, because the program considered here is free of branchings and loops. However, in contrast to execution record for tree-based GP (Example 3.1), the order of instruction execution is this time unambiguously determined by the program.

Inspection of this execution record reveals that execution traces for different tests may reach the same state at some point of program execution; for instance $p^1(t_1) = p^1(t_3) = 00$ while $p^0(t_1) \neq p^0(t_3)$ (where we abuse the notation and write $p(t)$ for an application of program p to the *input* part of test t). We say that such traces have *merged*.   ■

merging
of
execution
traces

Merging of execution traces is common and due to the fact that instructions often realize many-to-one computations. Examples abound: elementary arithmetic operators map infinitely many different pairs of arguments to the same outcome. The *sign* function accepts any real number but produces only three distinct outputs: $sgn : \mathbb{R} \to \{-1, 0, 1\}$. This applies also to more sophisticated concepts that may form single instructions in higher-level and/or domain-specific languages. A program/instruction that tests its argument for primality has the signature $\mathbb{N} \to \mathbb{B}$. A program/instruction that calculates the greatest common denominator (GCD) for a pair of numbers yields the same outcome for infinitely many pairs of arguments (e.g., $GCD(6,9) = 3 = GCD(21,24)$). A sorting program/instruction returns the same sorted list of length k for $k!$ different input lists. Many fundamental mathematical concepts (and henceforth many elementary instructions in programming languages) are many-to-one. Those that are one-to-one (and thus implement *injections*) are few and far between; prime factorization is a well-known example of the latter.

Therefore, some of the execution traces may merge while some may not. Crucially, once some traces have merged, they *cannot diverge anymore* in the course of further execution, because the environment fully determines the further course of execution. That would be possible only for nondeterministic instructions (which we excluded in the very beginning of this book).

Whether merging of particular traces is desirable or not becomes clear only when a program is confronted with a specific program synthesis task. Assume that the program in Fig. 6.1a is evaluated in the task where the target, i.e. vector of desired outputs, is $t^* = (0,0,1,1)$. The corresponding actual outputs of the program are $(0,0,0,1)$, i.e. the marked in bold elements of the column marked by $p^\$$ Fig. 6.1b, as we assumed that r_1 is designated to hold the effective output. These vectors differ at the first and third position, which can be explained in terms of execution traces: the traces for t_1 and t_3 have merged after executing the first instruction (column p^1), so it was clear already at that point that they must end up in the same final when reaching $p^\$$. Given the different desired values for tests t_1 and t_3, it was inevitable for the program to fail one of them.

This scenario exemplifies the key observation that gave rise to the approach we proposed in [100]: whether a particular merger of traces is desirable or not can be stated while a program is still in the course of execution, i.e. at *intermediate execution states* encountered before reaching $p^\$$. In particular:

intermediate execution states

Case 1: If different program inputs should be mapped to *different* program outputs, then merging the corresponding traces is undesirable. A program that does so loses the information necessary to distinguish between these inputs, and necessarily fails some of those tests. Merging of the traces for t_1 and t_3 in Example 6.1 represents to this case.

Case 2: By contrast, if different program inputs are supposed to be mapped to *the same* output, then merging the corresponding traces is desirable. Reaching the same execution state for these inputs ensures producing ultimately the same output, because once traces merge, they cannot diverge anymore. Merging of the traces for t_1 and t_2 at $p^\$$ in Fig. 6.1 is desirable in this sense.

The conventional objective function (1.7) indirectly depends on trace mergers because they influence program output. However, detecting mergers at earlier execution stages may help discovering programs that reach the *potentially useful* intermediate execution states, even when the actual program output is utterly wrong. A program featuring such states has the potential of being improved by the forthcoming moves of a synthesis algorithm (e.g., by augmenting it with a proper program suffix). In the next section, we present an alternative evaluation function that rewards programs according to the two type of trace mergers mentioned above.

6.2 Trace consistency measure

In the following, we design an evaluation function that quantifies the occurrence of correct and incorrect mergers in terms of information theory.

Consider a random variable X_k corresponding to the kth column in an execution record (3.5), i.e. characterizing the states generated by kth program instruction for particular tests. The *information content* of X_k is the entropy of its probability distribution. When expressed in bits, it amounts to:

infor- mation content of execution state

$$H(X_k) = -\sum \Pr(X_k) \log_2 \Pr(X_k). \tag{6.1}$$

For instance, for the column corresponding to p^0 in Fig. 6.1

$$E(X_0) = -2\frac{1}{2}\log_2\frac{1}{2} = 1, \tag{6.2}$$

and for $p^\$$

$$E(X_\$) = -\frac{1}{4}\log_2\frac{1}{4} - \frac{3}{4}\log_2\frac{3}{4} \approx 0.81. \tag{6.3}$$

Whenever traces merge at the kth instruction of a program, i.e. when passing from the $(k-1)$th to the kth column of an execution record, this is reflected in differences between probability distributions of corresponding random variables X_{k-1} and X_k. Subject to the desired outputs, those changes may be undesirable (Case 1 in the previous section) or desirable (Case 2). To assess the utility of these cases, we use *conditional entropy* $H(Y|X) = -\sum \Pr(Y|X) \log_2 \Pr(Y|X)$, with the dependent random variable Y associated with the desired output. We define two quantities that correspond to Cases 1 and 2 respectively:

1. $H(Y|X_k)$, i.e. the amount of information that desired output Y adds to X_k. In particular, if $H(Y|X_k) > 0$, then X_k alone is not sufficient to predict the value of Y.

uivalence
ass of
ecution
ates
2. $H(X_k|Y)$, the amount of information that X_k adds to Y. Large values of $H(X_k|Y)$ indicate that X_k partitions the set of tests into many *equivalence classes*.

Every time the traces for two or more tests merge between the kth and $(k + 1)$th execution step (columns of the execution record), either the former term increases ($H(Y|X_k) > H(Y|S_{k+1})$) or the latter term drops ($H(X_k|Y) > H(X_{k+1}|Y)$). Both $H(Y|X_k)$ and $H(X_k|Y)$ attain zero if and only if X_k is perfectly *consistent* with Y, i.e. $s_i^k = s_j^k \iff y_i = y_j$.

Following this observation, we devise an evaluation function based on the sum of the above terms. The $f_{TC}(p)$ of a program p is the minimal *two-way conditional entropy* defined above, calculated over all variables X_k, i.e.

trace consistency

$$f_{TC}(p) = \min_k H(Y|X_k(p)) + H(X_k(p)|Y), \qquad (6.4)$$

where $X_k(p)$ is the random variable associated with the kth column in the execution record of program p. Lower values of f_{TC} indicate merging of traces that is more consistent with Y and thus more desired; therefore, f_{TC} is to be minimized. By using the minimum operator for aggregation over program execution steps, $f_{TC}(p)$ rewards p for the part of its behavior that is most consistent with the desired output. This is intended to promote the programs that feature code fragments (subprograms) that can prove useful in new programs[1].

Example 6.2. Consider a program synthesis task with five tests, and an execution record shown in Table 6.1. The actual program that produced this record is not important here. It is sufficient for us to know that an execution state comprises three single-bit registers, and two of them are interpreted as program output (column Y of the table). However, it is also irrelevant which of the registers are interpreted as program output, as the trace consistency measure f_{TC} never directly compares execution states with desired output; it cares only about the information content associated with particular random variables corresponding to the columns of execution records.

The lower rows of the table present the conditional entropy for consecutive columns. As the traces merge, the corresponding random variables X_k carry less and less information, and so the entropy of Y conditioned on X_k may only grow. For instance, the states 001 and 010 for the tests t_2 and t_3 in X_0

---

[1] Note that the term minimized in (6.4) can be alternatively expressed as $H(Y, X_k(p)) - I(Y; X_k(p))$, where $H(Y, X_k(p))$ stands for joint entropy and $I(Y; X_k(p))$ is the mutual information. Minimization of (6.4) is thus not equivalent to maximization of mutual information only, as also $H(Y, X_k(p))$ may vary between columns of an execution record.

Table 6.1: Exemplary calculation of consistency measure for an execution record of a program applied to five tests, each comprising three Boolean input variables and two Boolean output variables. The minimum of two-way conditional entropy $H(Y|X_k) + H(X_k|Y)$ over the columns of execution record, marked in bold, is the trace consistency f_{TC} of this program. For readability, the unique intermediate states are marked in colors; note however that the measures considered here never compare states across columns of an execution record, so what matters is only the colors within particular columns.

Test	p^0	p^1	p^2	Y
t_1	000	000	010	00
t_2	001	001	001	10
t_3	010	001	001	10
t_4	011	010	010	10
t_5	100	011	010	11
$H(Y|X_k)$	0	0	0.95	
$H(X_k|Y)$	0.95	0.55	0.55	
$H(Y|X_k) + H(X_k|Y)$	0.95	**0.55**	1.50	

merge into the state 001 in X_1; this however does not increase $H(Y|X_k)$ because the desired output for these tests is the same (10). On the other hand, when the traces for t_1, t_4 and t_5 merge in X_2, this leads to an increase of $H(Y|X_k)$, because these three tests have different desired outputs Y, so this merging is undesirable.

Conversely, $H(X_k|Y)$ cannot grow with consecutive columns. When it drops in p^1, it is because the traces for t_2 and t_3 have merged, which is desirable. On the other hand, it does not drop between p^1 and p^2, because it is where an undesirable mering of t_1, t_4, and t_5 takes place.

In an ideal case, $H(Y|X_k) = H(X_k|Y) = 0$, i.e. neither the variable associated with kth intermediate execution state adds any information to the desired output, nor the reverse. This would happen if 001 and 010 collapsed into a single state in X_2 and t_4 did not merge with t_5 there, as a result of which X_2 would be perfectly consistent with Y. ∎

Note that attaining $f_{TC}(p) = 0$ does not necessarily mean that p features a subprogram that solves a given program synthesis task, i.e. produces intermediate results that are equal to the desired output. This would only signal that one of X_ks is perfectly *consistent* with the desired output, i.e. that there exists a one-to-one mapping between the intermediate memory states at the kth step of program execution and the desired outputs. For f_{TC}, it is only the *cquivalcncc* of states that matters, not the states themselves. In this sense, f_{TC} can be said to be more lenient than f_o. This also facilitates technical implementation: one can be oblivious about the interpretation of data that form the states; knowing how to test them for equality is sufficient. On the other hand, we cannot verify program correctness by asking whether $f_{\mathrm{TC}}(p) = 0$, which was possible for f_o. However, this is consistent with Sect. 1.3, where we clearly delineated the correctness predicate *Correct* from an evaluation function. This separation is also consistent with the concept of search driver to be presented in Chap. 9.

Trace consistency measure f_{TC} reveals thus certain 'internal qualities' of candidate programs that the conventional evaluation function is oblivious to. It forms yet another behavioral characteristic that may be used as an alternative (or a supplement) to the conventional objective function. The experiment in Chap. 10 verifies this hypothesis.

6.3 Trace consistency for non-linear programs

The line of reasoning in the previous sections is valid only for strictly sequential programs where execution states reflect the entire execution environment, as in the formal definition of execution record (3.5). In such environments, future computation is fully determined by the current state and traces cannot diverge anymore once they have merged.

In tree-based GP, the arguably most popular genre of GP, program execution is not sequential nor does it maintain global memory. An intermediate outcome at a given instruction (tree node) reflects only the computation carried out by the subtree rooted in that node. In the absence of side effects (i.e. when instructions are pure functions) the order of execution is only partially determined by the structure of a program (see Example 3.1). There is no obvious, 'natural' succession of instructions, and instructions of can be executed in various order without affecting the final semantics. Talking about 'further course of computation' does not make sense in such a context.

Therefore, the definitive statements about merging and divergence of execution states need to be relaxed for tree-based GP (and other genres of GP where programs are not strictly sequential), which we illustrate with the following example.

Example 6.3. Consider the purely functional tree-based GP program (AND (AND x_1 x_2) x_2) applied to input composed of two Boolean values x_1 and x_2. Figure 6.2 presents the execution record of this program for tests that enumerate all combinations of inputs (x_1, x_2). Each column in the record corresponds to an instruction in the program, executed in bottom-up, left-to-right order. An execution state is the value returned by the just executed expression, and depends only on the corresponding subtree. Consider the column p^3 of execution record, corresponding to the subtree (AND x_1 x_2).

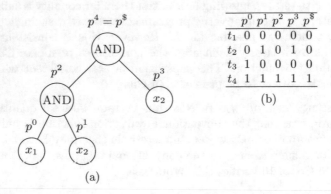

	p^0	p^1	p^2	p^3	$p^\$$
t_1	0	0	0	0	0
t_2	0	1	0	1	0
t_3	1	0	0	0	0
t_4	1	1	1	1	1

(b)

(a)

Fig. 6.2: A tree-based, side effect-free GP program (a) and its execution record (b), built assuming bottom-to-top, left-to-right order of executing instructions. Execution traces for t_1 and t_2 merge in p^2, but diverge later in p^3, because under this program representation execution states do not fully determine the course of further program execution.

Because the yet-to-be-conducted computation depends on program input (the rightmost node x_2), the traces that met at p^3 can still diverge. This is what we observe for t_1 and t_2 in Fig. 6.2: their traces merge when reaching p^2 but diverge again in p^3. However, this divergence is only an artifact of adopting a particular order of executing instructions.   ∎

local
execution
state Execution states in tree-based GP are thus in a sense *local*. While merging traces for a subset of tests means that a sequential program is destined to end up with the same output, for tree-based GP that still depends on the course of execution of the remaining part of a program. A merger at some point of execution *lowers* the odds for the engaged traces diverging in the remaining part of execution, but does not exclude it. The chance of such a divergence decreases as the end of the program is approached. The traces that merge deeper in a program tree have more chance to diverge than those that happen closer to the root node.

This relaxation concerns Case 1 in Sect. 6.1, i.e. merging the traces that should not merge for a given subset of tests. An analogous argument applies also to Case 2. Traces that should have merged but did not at an intermediate execution stage can be still merged – even the very last instruction of a program may fix this. However, the longer a program postpones such a desired merger, the less likely it is that the remaining instructions will do so.

Thus, although for tree-based GP merging and diverging of traces do not determine consistency with program output as directly as it was for the strictly sequential GP, they can be still considered as telltales of the prospective performance of a program. This makes the trace consistency measure

f_{TC} (6.4) a viable evaluation function also for tree-based GP. Indeed, in [100] we demonstrated experimentally that tree-based GP driven by f_{TC} outperforms the conventional objective function f_o on popular GP benchmarks. Explanation of that result goes along the rationale presented in this chapter. A subprogram p' in a program p that behaves highly consistently with the desired output is prospectively valuable. Application of a search operator to p has a non-zero chance to produce a well-performing program by combining p' with an appropriate 'suffix'. The conventional evaluation function is oblivious to the internals of program execution and thus unable to promote such programs.

The incompatibility of non-sequential programs with sequential, tabular execution records points to possible generalizations of the latter. Ideally, a generalized execution record would reflect the causal relationships between program parts in a more direct way. Another future work may concern handling programs that feature loops or even recursion. That would however require a more sophisticated alignment of traces that that which follows directly from program syntax.

6.4 Summary

The trace consistency evaluation function f_{TC} is the first formalism presented in this book that depends on an entire execution record. By this virtue, it may reveal differences between evaluated programs that escape the attention of semantic GP methods (Chap. 5) and methods that behaviorally assess test difficulty using outputs produced by programs (Chap. 4). One may thus anticipate observable differences between f_{TC} and the other evaluation functions presented in this book, which we will verify experimentally in Chap. 10.

There are though still certain nuances in program execution that the consistency-based evaluation is oblivious to. In particular, f_{TC} cares only about the *equivalence* of states across traces. What matters is only whether traces reach identical execution states, not *what* those states actually represent. This is on one hand an elegant abstraction from the internals of execution state; on the other, one may suppose that the actual *content* of execution states can be more informative and allow designing more sophisticated evaluation functions. The approach presented in the following chapter follows this observation.

7

Pattern-guided program synthesis

The motivation behind analyzing consistency of execution traces with desired output in Chap. 6 was to identify and promote the programs that contain prospectively useful subprograms. The approach described in this chapter generalizes the trace consistency method in two respects. Firstly, we seek here for a more general *relatedness* between the intermediate execution states and the desired output, rather than for information-theoretic consistency. Secondly, while evaluation in the trace consistency method depends on a single stage of program execution that maximizes consistency (captured in a particular column of an execution record (6.4)), the evaluation functions proposed here depend on the entire execution record. In this way, the approach presented in this chapter, originally proposed in [101] and then extended in [96], looks for *patterns in program behaviors* that seem relevant for a given program synthesis task.

7.1 Motivation

The motivations for relying on general 'relatedness' between intermediate execution states and program output and for taking into account entire execution records can be illustrated with the following example.

Example 7.1. Assume the task is to synthesize a program that checks if a quadratic polynomial $ax^2 + bx + c$ has roots in the real domain. The input to the program is a triple of numbers (a, b, c) and the desired output is an appropriate Boolean value. Consider the candidate program for this task, which we present as a function in the Scala programming language in Fig. 7.1a to demonstrate that the formalism of execution record is applicable to conventional programming languages.

Fig. 7.1b shows the corresponding execution record for this program, composed of three traces. The program is clearly very close to being correct; its only deficiency is disregarding the `delta` variable. Fixing this is straightforward for a human programmer: the `return` statement in line 4 needs

© Springer International Publishing Switzerland 2016
K. Krawiec, *Behavioral Program Synthesis with Genetic Programming*,
Studies in Computational Intelligence 618,
DOI: 10.1007/978-3-319-27565-9_7

```scala
1  def hasRoots(a,b,c:Double):Boolean = {
2    val hasDegreeTwo = (a != 0)
3    val delta = b*b - 4*a*c
4    return hasDegreeTwo
5  }
```

(a)

	a b c	p^2	p^3	p^4
t_1	1 2 2	$true$	-4	$true$
t_2	-1 2 2	$true$	12	$true$
t_3	0 2 2	$false$	4	$false$

(b)

Fig. 7.1: (a) A partially correct program in Scala for determining if a quadratic polynomial $ax^2 + bx + c$ has real roots. (b) The execution record for this program for three exemplary inputs. The columns p^i of the execution record are numbered consistently with the line numbers in the program listing.

to be extended to `return hasDegreeTwo && (delta >= 0)`. The components required for expressing the target concept are calculated at intermediate execution stages in lines 2 and 3. However, this program fails to combine them in the right way.

The conventional objective function f_o would judge this program only by its output, which would be incorrect for many tests. The trace consistency evaluation function f_{TC} (6.4) from Chap. 6 should be able to notice that the intermediate execution states reached in lines 2 and 3 (i.e. the values of variables `hasDegreeTwo` and `delta` calculated there) have relatively high information-theoretic consistency with the desired output. Depending on the actual tests used, the two-way conditional entropy is likely to be minimized by one of the random variables X_k associated with these locations (6.4). However, only one of them will be ultimately reflected by f_{TC}. f_{TC} cannot take into account that it is the *combination* of these expressions (subprograms) that is particularly promising. As a consequence, the above program may attain the same value of f_{TC} as its close relatives, e.g., a program that misses line 3. ∎

The lesson learned from this example is that consistency of execution states with the desired output is only one type of potentially useful behaviors that can emerge in candidate programs synthesizes in an iterative framework like GP. There are other, more complex and more subtle behaviors that can be the telltales of prospective performance. The challenge lies in making a behavioral pattern program synthesis method capable of detecting such *behavioral patterns*.

A skilled human programmer may discover behavioral patterns and exploit them to design a program that meets the specification of a program synthesis task. Humans in general are known to be incredibly good at spotting patterns and thinking in patterns when solving all sorts of problems – it is not for no reason that they have been termed *informavores* [128]. A great deal of AI research is about modeling and mimicking such capabilities [53]. Moreover, humans can *anticipate* the patterns that are desirable in a given problem and often use domain and common sense knowledge for that sake.

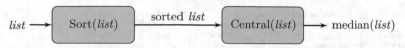

Fig. 7.2: The conceptual decomposition of a generic algorithm calculating the median of a list of numbers.

Example 7.2. Consider the task of synthesizing a program that calculates the median of a list of numbers (Fig. 7.2). The background knowledge tells us that a reasonable first stage of solving this task is to sort the list. Once the input list is sorted, its median is the central element (for odd-length lists, or the mean of two central elements for lists of even length). In terms of execution records, reaching an intermediate execution state that contains the sorted elements of the list is desirable for this task.   ∎

Following the research we reported in [101, 96], we propose to consider synthesis approaches that mimic human programmers in detecting the potentially useful behavioral patterns and reward the programs accordingly with evaluations. Hereafter, we use the term PANGEA (PAtterN Guided Evolutionary Algorithms) to describe this class of approaches. The particular representative of this class described in this chapter employs knowledge discovery algorithms to search for behavioral patterns. Such methods should in particular be able to notice the promising co-occurence of intermediate execution states in Example 7.1 and their joint 'relatedness' to the desired output. If this approach is able to reveal meaningful dependencies between partial outcomes and the desired output, we may hope to thereby promote programs with the potential to produce good results in future, even if at the moment of evaluation the output they produce is incorrect.

7.2 Discovering patterns in program behavior

Example 7.1 shed some light on the notion of relatedness signaled in the beginning of this chapter. The concepts `hasDegreeTwo` and `delta` are related there to the target concept `hasRoots` in the sense that they can be easily combined into a correct program. The trace consistency method in Sect. 6.2 could be extended so as to capture such co-occurrences of useful concepts (by, e.g., examining multiple conditioning variables X_k simultaneously). There is however a more natural way, which also opens the door to a wider class of behavioral patterns. Synthesis of an entity that captures a given target concept using some elementary features can be phrased as supervised *learning from examples*, arguably the most intensely studied paradigm of machine learning (ML).

learning
from
examples

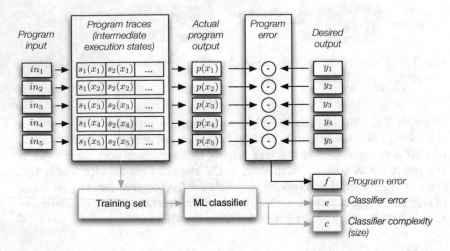

Fig. 7.3: The workflow of Pattern-Guided Evolutionary Algorithm. (PANGEA)

The ML perspective on behavioral program synthesis originates in the observation that an execution trace bears certain similarity to an example in ML (cf. Fig. 3.2). Recall that a program p applied to an input in reveals its behavior with the execution trace $(p^0(in), \ldots, p^\$(in))$ (3.5). If an execution record resulting from applying p to all tests in T is aligned, i.e. the states in particular traces correspond to each other, the columns X_k of the record can be likened to *attributes* in ML. The desired program output *out* corresponds in this context to the desired response of a classifier[1]. And crucially, an ML induction algorithm (*inducer* for short), given a set of such examples, can be used to learn a classifier that predicts *out* based on the attributes X_k describing execution traces.

Similarly to the trace consistency method and test difficulty-based methods, PANGEA defines its own evaluation function, which is intended to replace or augment the conventional objective function. The method proceeds in the following steps, depicted in Fig. 7.3:

1. An execution record is built by running the program on the tests.
2. The execution record is transformed into an ML dataset D.
3. An ML induction algorithm is applied to D, resulting in a classifier C.
4. Program evaluation is calculated from C.

Construction of an execution record in Step 1 has been covered in Sect. 3.1. Steps 2–4 are detailed in the following subsections.

---

[1] Depending on the domain of *out*, the ML task in question is classification or regression. Although pattern-guided program synthesis can be applied to both, for brevity we use the term 'classification'.

7.2.1 Transforming an execution record into an ML dataset

An execution record in its original form (3.3) is not suitable to learn from using the conventional supervised ML techniques. Execution states can represent any data (e.g., single values in Fig. 6.2 or pairs of bits in Fig. 6.1), while ML algorithms typically expect metric or nominal attributes. The record also does not have an explicitly defined dependent variable, which is indispensable for supervised learning. In principle, one could overcome these difficulties and devise a specialized ML algorithm to learn directly from an execution record. We find it however more reasonable to transform an execution record into a conventional dataset and so make it amenable to a multitude of ML inducers that may represent hypotheses in various ways (e.g., as decision rules, decision trees, or neural networks).

An execution record of m traces and n columns is transformed into a machine learning dataset D composed of m examples and $n' \geq n$ attributes. Each trace (row) in the execution record corresponds to an example in D, and every column gives rise to one or more attributes (a.k.a. *features*). An attribute is thus a complete or partial image of executing environment at particular stage of program execution. The attributes are derived from the states in a way that depends on GP genre. In tree-based GP, there are no side effects and every instruction (a node in program tree) returns a single value. If those values represent a simple type, which we assume in this book, they directly form the corresponding ML attribute. The type of the attribute is set accordingly to the type of data returned by an instruction: floating-point values give rise to continuous attributes, discrete values (booleans, enumerations, etc.) to nominal attributes or ordinal attributes. For instance, the execution record for the tree-GP program in Fig. 6.2 would be composed of five nominal attributes corresponding to particular nodes in the program tree.

In non-tree GP paradigms (or more precisely the non-functional ones, i.e. involving side effects), the course of the above transformation depends on the particular form of execution record. In Sect. 3.2, we suggested to build execution records in a differential fashion, i.e. so that the consecutive columns reflect only the *changes* in an execution environment rather than the entirety of them. This guarantees the columns to be mutually non-redundant, which comes in particularly handy in the context of ML, where redundant attributes are a waste of resources in the best case (and can reduce classifier performance in some scenarios).

Consider the simple, non-differential execution record shown in Fig. 7.4a, reproduced for convenience from $6.1b$, where it was generated by a linear GP program. Every column of that record would give rise to two nominal ML attributes corresponding to the registers r_1 and r_2 that form the execution environment. That execution record would thus result in an ML dataset comprising six attributes. The differential execution record for the same program

p^0 p^1 $p^\$$
t_1 00 00 00
t_2 01 01 00
t_3 10 00 00
t_4 11 11 11

(a)

p^0 p^1 $p^\$$
t_1 00 00 00
t_2 01 01 00
t_3 10 00 00
t_4 11 11 11

(b)

Fig. 7.4: The execution record (a) from Fig. 6.1 and its differential version (b). The elements of execution state that do not change and disappear from the differential record are grayed out. These execution records can be directly mapped to ML datasets that feature respectively six and four attributes.

is shown in Fig. 7.4b: the grayed-out are omitted in that record, as they have not been affected by the corresponding instructions. In consequence, the resulting ML dataset would comprise four attributes only.

Analogous simplifications when building an ML dataset from an execution record hold for the Push programming language [170], which is also non-functional and involves side effects that manifest in changing the states of data and code stacks that form the execution environment there. However, most Push instructions modify only selected elements of specific stacks. To avoid redundancy, it is thus desirable to define the corresponding ML attributes, e.g., the top element of the stack affected by the just executed instruction. For instance, the *.INTEGER instruction pops two elements from the top of the integer stack, multiplies them, and pushes the result back on top of the integer stack. That result becomes the value of the associated attribute. An instruction producing multiple output values (e.g., swapping two topmost elements on a stack) would by the same principle give rise to two attributes. In [101], we successfully applied such proceeding when applying PANGEA to programs evolved in Push. The design of ML attributes is related to the incremental nature of practical implementations of execution records, which we discussed in Sect. 3.2.

To enable discovery of dependencies between execution traces and the target behavior, the resulting dataset D needs to define a supervised learning problem. To this end, we augment it with a *decision attribute* (dependent variable) that reflects the desired output. The type $\mathcal{O}$ associated with desired output is determined by a given program synthesis task (Sect. 1.2). In this book we focus on tasks typically considered in GP, where the desired output is a scalar. In such cases, the corresponding decision attribute in D is identical to it. When the desired output is nominal, so is the decision attribute in D, and D defines a classification task. If the desired output is continuous, D defines a regression task. A program synthesis task with desired output forming a compound entity (e.g., list) would require mapping into several decision attributes.

In summary, the dataset serves as a representation layer for the execution record, an 'adapter' between it and an ML induction algorithm. For tree-based GP and simple data types (Booleans, integers, reals) it is essentially transparent, replicating one-to-one the execution record and the desired outputs.

7.2.2 Classifier induction

The outcome of the previous stage is a conventional ML dataset D of m examples, each described by n' attributes (where $n \leq n'$ is the number of columns of an execution record) and a decision attribute. The objective of PANGEA is to assess how useful are the attributes in D for predicting the dependent attribute, i.e. the desired output of the program. We achieve this by applying an induction algorithm (inducer) to D and so obtaining a trained classifier. The induced classifier models the dependencies between the intermediate execution steps and the desired output. In doing so, it implicitly detects patterns in execution traces that are related to the target of the program synthesis task.

Subject to certain constraints discussed in the next section, many types of inducers can be used here to train a classifier. In the experiments reported in Chap. 10 PANGEA employs a decision tree induction algorithm, so we use it also in the example that follows.

Example 7.3. Figure 7.5 presents the above process for the program reproduced from Figs. 3.1 and 5.1 and an integer-valued symbolic regression task composed of four tests and two input variables. The colored lists below instructions in the program tree show the intermediate execution states produced for the four tests. When gathered together, they form the execution record (not shown here) comprising seven columns, i.e. as many as there are nodes in the program tree. Among them there are columns that reproduce the input variables (x_1 and x_2) and columns that reflect the intermediate execution states (the left and the right subtree of the root node). The execution record is subsequently transformed into an ML dataset featuring four attributes (x_1 to x_4), where the redundant columns (multiple occurrences of x_1) are removed. The blue column corresponding to program output is also discarded in order to focus the evaluation on the intermediate execution states. Given this training set, a decision tree induction algorithm produces the presented classifier, interpreting the attributes as nominal variables. The resulting decision tree is characterized by $f_e = 0$ and $f_c = 5$; these two values form bi-objective characterization of the program. The value of the conventional objective function f_o (here: city-block metric) for this program amounts to 10. ∎

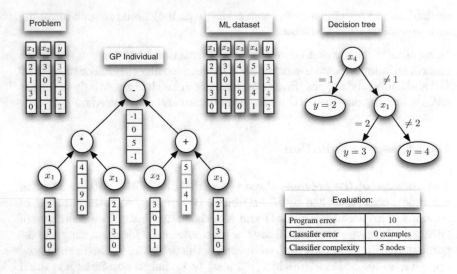

Fig. 7.5: Calculation of pattern-guided evaluation for a program reproduced from Figs. 3.1 and 5.1. See Example 7.3 for detailed explanation.

7.2.3 Evaluation functions

Example 7.3 illustrates that the classifier maps the intermediate execution states (capture in the attributes of ML dataset) onto the desired output of the program. In a sense, it attempts to complement program's capability for solving the problem (i.e. producing the desired output value). Crucially, if the traces feature regularities that are relevant for predicting the desired output, then the induction algorithm should be able to build a classifier that is (i) compact and (ii) commits relatively few classification errors. These aspects are strongly related to each other, which we illustrate in the following.

Consider first a correct program p. p solves the task, i.e. produces the desired output $out_i = p(in_i)$ for all tests $(in_i, out_i) \in T$. Since each trace ends with a final execution state, the last attribute $X_\$$ in D reflects the desired output (for side effect-free programs and simple data types $X_\$$ is actually equivalent to desired output). In such a case, the inducer should be able to produce a classifier that uses only $X_\$$. For instance, for decision tree inducers, the resulting tree could be composed of a single decision node involving $X_\$$ and k leaves corresponding to the k unique values of desired output. Such a decision tree is thus quite compact and commits no classification errors.

Now consider an incorrect program such that its output diverges so much from the desired output that the corresponding attribute $X_\$$ is useless for prediction. In such a case, it is likely for the induced classifier to ignore it and rely the other attributes, derived from the intermediate execution states. Each such attribute individually has usually limited predictive power. In consequence, the resulting classifier needs to rely on many such attributes and thus be quite complex. In the case of decision trees, the tree will feature many decision nodes. In general, the size and predictive accuracy of the classifier depend on the degree to which the intermediate states *relate* to the desired output.[2]

These examples illustrate that complexity and predictive capability of a classifier are strongly related, a fact that is well known in ML. According to the *Minimum Description Length* principle (MDL [152]), the complexity of the mapping from the space of attributes onto the dependent variable may be expressed by summing the encoding lengths of the classifier (the 'rule') and of the erroneously classified examples (the exceptions from the rule). However, practice shows that plain sum of these two terms rarely works in practice and they need to be carefully weighed, which can be challenging. For this reason and driven by the motivations discussed later in Chap. 9, we define in PANGEA two separate evaluation functions. The first of them is the *classification error*

Minimum Descrip- tion Length

classifi- cation error

$$f_e(p) = |\{(in_i, out_i) \in T : C(D_i) \neq out_i\}|, \qquad (7.1)$$

where C is the classifier induced from the training set D, D_i is the ith example in the training set D, and $C(D_i)$ denotes C'a prediction for example D_i. The second evaluation function is *classifier complexity*

classifier complex- ity

$$f_c(p) = |C|, \qquad (7.2)$$

where $|C|$ is the complexity of the classifier C, calculated accordingly to classifier representation. For decision trees, it is simply the number of tree nodes.

As it follows from the earlier argument, neither $f_e(p)$ nor $f_c(p)$ alone captures the relatedness of attributes to the desired output. In consequence, we usually use them alongside each other.

Example 7.4. This example continues Example 7.3 illustrated in Fig. 7.5. The decision tree induced there from the execution record of program p has five nodes, so $f_c(p) = 5$. When tested on the tests from Fig. 7.5a, that classifier commits zero error, therefore $f_e(p) = 0$.

Now consider another program $p_2 = (* \, x_1 \, x_1)$. Because this program is a subprogram of p (Fig. 7.5b), the resulting dataset comprises only the first two attributes shown in Fig. 7.5c. The ML inducer can build a zero-error decision tree from these data, which is possible but requires seven nodes. In such

---

[2] For clarity, in this explanation we assume classifiers that explicitly select attributes, like decision trees or decision rules. Non-symbolic classifiers (like SVMs or neural nets) would require more elaborate means to demarcate the used/selected attributes from the unused ones; see Sect. 7.3.

a case, $f_c(p_2) = 7$ and $f_e(p_2) = 0$. Of course, particular ML inducers vary in the way they prioritize classifier complexity and error, so the characteristics of the underlying ML algorithm determines in part the characteristics of f_c and f_e.   ∎

Similarly to f_{TC} in the trace consistency approach (Chap. 6), f_e and f_c assess program's 'prospective' capabilities. If a program calculates intermediate results (i.e. feature subprograms) that relate to the target concept defined by the program synthesis task, the resulting classifier is likely to be simpler and/or commit fewer errors. This will happen even if the actual output of the program is useless for the task being solved. However, while f_{TC} seeks a *single* behaviorally most consistent subprogram, an inducer can spot co-occurrence of several subprograms which, when combined together in the classifier, allow attaining low classification error.

Execution records also reflect input data fed into a program, and this manifests in the attributes that reproduce the input variables (like X_1 and X_2 in Fig. 7.5). As we assumed the tests in T to be coherent (Sect. 1.3), the sought mapping from program inputs to the desired outputs in T is a function. Therefore, if classifier representation is expressive enough, there always exists a classifier that maps the attributes to the desired output at no error ($f_e = 0$). For instance, for any coherent supervised classification dataset composed of nominal attributes there exists a decision tree that commits zero error. One might thus question the rationale behind PANGEA. Note however that realizing a direct mapping from program input to program output without the help of attributes based on intermediate execution states will usually require a complex classifier. The absence of useful subprograms is thus penalized by large f_c. Once such subprograms emerge in candidate programs in the further course of iterative synthesis process, an induction algorithm may prefer to use them because they facilitate construction of simpler classifiers.

Most inducers exhibit such *inductive bias*, which has its roots in Occam's Razor and is intended to lower the likelihood of overfitting. In PANGEA, that bias determines the way an inducer handles the trade-off between f_e and f_c. Alternatively, one may consider an abridged variant of execution record that does not include the attributes derived from input data or even covers only selected stages of program execution, for instance only a few last executed instructions, as in [101].

It may be also worth realizing that most ML inducers are heuristic algorithms that do not guarantee producing an optimal classifier for a given input data (here: transformed execution record). For instance, decision tree inducers like C4.5 [150] search greedily the space of possible tree designs and may yield a classifier that is suboptimal with respect to size, classification accuracy, or both. Therefore, the values of f_e and f_c used in PANGEA only approximately characterize the evaluated programs. However, we do not find this critical, given that program synthesis as conducted by GP is heuristic and stochastic in the first place.

7.3 Discussion and related concepts

PANGEA conducts the four-stage evaluation process described above for each candidate program individually, in abstraction from the other programs in a working population P. As the programs in P may vary in length, the training sets D constructed from them vary in the number of attributes. Nevertheless, if the same induction algorithm is used, the resulting evaluations on f_e and f_c are comparable, because they characterize the classifiers that solve the same underlying problem. The decision attribute, derived from the desired output of program synthesis task, is the same, and so is the input data that gives rise to program traces and attributes. Put in ML terms, PANGEA uses an ML inducer to evaluate the utility of various data preprocessors, implemented by subprograms.

Obviously, transformation of an execution record into an ML dataset (Sect. 7.2.1) can be made more sophisticated. That would be necessary if execution states represented not elementary but compound data types, or when representation bias or inductive bias of an inducer prevented it from capturing patterns that were essential for solving the task. Yet another motivation is to allow discovery of higher-order patterns that are unobservable when each attribute reflects a single execution state. Looking for some form of analogies [53] between execution traces is an appealing option here. It is interesting to note that similar motivation propelled the research on *feature construction* in ML (see review in [83]).

In the past, GP has been often combined with ML in a similar modus operandi to synthesize features. Such hybrids were routinely used to solve ML tasks and proved particularly effective when applied to challenging problems in pattern recognition and computer vision (see, e.g., [91, 83, 11, 89, 178] and [92] for a review). In those studies however, feature definitions evolved with GP form an inherent part of the target ML system. In PANGEA, the situation is in a sense reversed: an ML inducer is at the service of program synthesis, and its usage is transient and limited to the evaluation process. The classifier is discarded once f_e and f_c have been calculated.

feature synthesis

Classifier complexity f_e is easy to compute for decision trees, where it boils down to counting the number of tree nodes. This is one of the reasons we adopted decision trees in the above examples and in the experimental part of this book. Nevertheless, there exists some means of expressing complexity for virtually all classifiers (models). For instance, in [3] we used the measures based on Akaike Information Criterion [141] and proposed by [172] to characterize the complexity of regression models expressed as vectors of real parameters.

The MDL principle that helps understanding the trade-off between f_e and f_c was used in GP in numerous occasions. In most such cases, it played a similar role to other machine learning techniques, i.e. as a means of controlling the trade-off between model complexity (sometimes referred to as *parsimony* in a GP context) and accuracy. In this spirit, Iba et al. [61] used it to prevent bloat in GP by designing an MDL-based evaluation function that took into account the error committed by an individual as well as the size of a program. A few later studies followed this research direction (see, e.g., [197]).

7.4 Summary

Among the methods presented in this book, the pattern-guided evaluation functions go the furthest in harvesting information about program behavior. The types of behavioral patterns PANGEA can detect is in principle limited only by the expressibility and the biases of the involved ML inducer. A typical inducer is able to, among others, detect the consistency of a single attribute with the desired output, which was the feature offered by trace consistency evaluation function f_{TC} in Chap. 6. We may claim thus that the class of behavioral patterns detectable by PANGEA forms a superclass of the patterns detectable by f_{TC}.

Nevertheless, f_e and f_c obviously do not reflect all details of program execution. Neither they nor f_{TC} ultimately solve the evaluation bottleneck problem. In place of a single scalar objective to drive a search, they present us with two objectives, still very little compared to the abundance of information available in an execution record. Thus, the door to even deeper behavioral evaluation remains open, which we touch upon in Chap. 11. Nevertheless, the experimental evidence in [101, 96] shows that even this modest widening of evaluation bottleneck immensely boosts the likelihood of synthesizing correct programs using compared to conventional GP. In Chap. 10, we will corroborate those results within a new experimental framework.

The common feature of the trace consistency approach and PANGEA is that they identify the columns in an execution record that are essential for the resulting evaluation. In the former, it is the column that minimizes the two-way entropy (6.4). In PANGEA, it is the columns that judged as pertinent to the desired output by an ML algorithm. In both approaches, such columns indicate specific *locations* in program code or, as we phrased this in this chapter, subprograms. One may wonder whether this additional information can be exploited for the sake of making program synthesis more effective. In the next chapter, we present an approach that originates in this observation.

8

Behavioral code reuse

To this point, our attempts to widen the evaluation bottleneck focused on defining alternative evaluation functions, which we conceptualize as *search drivers* in Chap. 9. However, an analysis of an execution record (Chap. 3), whether conducted with information-theoretic measures (Chap. 6) or machine learning algorithms (Chap. 7), also reveals information about the qualities of particular *components* of candidate solutions, i.e. subprograms. In this chapter, we elaborate on this observation and propose a means to harness its potential. We show how the detailed information available in an execution record together with behavioral evaluation enables (i) identification of useful components of programs, which can be then (ii) archived and (iii) reused by search operators. The following sections detail these stages as realized in [96].

8.1 Identification of useful subprograms

Many machine learning classifiers perform internal feature selection by deciding which attributes to use to construct a classifier. Some of them explicitly reveal that information. This in particular applies to the classifiers that represent their hypotheses (models) symbolically, among others decision trees used in the previous section. For instance, the decision tree in Fig. 7.5 engages only two attributes (x_1 and x_4) of the four available in the dataset. Similar transparency holds for rule-based classifiers and other more or less 'whitebox' symbolic representations like Bayesian networks. Nevertheless, the nonsymbolic representations do not preclude such possibility: the weights of a neural network can be for instance used to quantify the relative importance of attributes (or even construct a corresponding symbolic representation [29]), and statistical tools exist to assess the importance of attributes in regression models (see e.g., [3]). In summary, almost every type of classifier can be used to estimate the *relevance* of attributes in a given supervised learning task.

© Springer International Publishing Switzerland 2016
K. Krawiec, *Behavioral Program Synthesis with Genetic Programming*,
Studies in Computational Intelligence 618,
DOI: 10.1007/978-3-319-27565-9_8

In PANGEA (Chap. 7), behavioral evaluation of a program results in an ML classifier that serves as a basis for calculating behavioral evaluation functions. The classifier is induced from a dataset built upon an execution record. The attributes in that dataset correspond to columns in the original execution record, which in turn point to concrete instructions in the program. This chain of dependencies leads thus to instructions that are judged as most promising by an ML inducer, i.e. producing intermediate results that relate to the desired output. Behavioral evaluation can be thus seen as a means of solving the *credit assignment* problem [130]: deciding which components of a compound candidate solution are responsible for its overall performance and how to distribute the total reward between them.

credit
assign-
ment

On the face of it, considering credit assignment in the context of program synthesis is disputable. Intricate interactions between instructions make it difficult to treat programs as anything but monoliths (c.f. Sect. 1.4). For instance, how much credit should the second line of the program in Fig. 7.1a receive in return for the overall performance of this program (which, by the way, is incorrect as a whole, making this question even harder)?

This question is difficult if not void when asked at the level of source code. However, considerations about credit assignment become appropriate on higher abstraction levels involving *structures* of code, i.e. blocks, procedures, or in general *modules*. Because modules typically have well-defined interfaces with the other components (e.g., function signatures) and certain roles in the context of entire program, one might quantify the degree to which a given instance of a module meets the expectations of the context. Put in terms of studies on *modularity*, such structures are *nearly-decomposable* but usually not *separable* (see [163, 190] and Sect. 11.1).

sub-
program

In tree-based GP, the natural interpretation of such a structure is a subtree in program tree, which we refer to in the following as *subprogram*. If a particular programming language is functional, there is no global state and subprograms perform independent computation. In such a case, the classifier in PANGEA, by referring to a particular column in an execution record, indicates usefulness of the subprogram rooted in corresponding node of program tree. As a side effect of program evaluation, we obtain a subset of useful subprograms. PANGEA is not the only approach in this book where evaluation can adopt this role of 'subprogram provider'. The trace consistency evaluation function f_{TC} in Chap. 6 has similar capability: formula (6.4) seeks the column in an execution record that minimizes the two-way conditional entropy, and that column pinpoints the subprogram that behaves most consistently with the desired output.

Given a program p, the subprograms identified by a classifier C in p will be denoted by $P_C(p)$. We allow $P_C(p) = \emptyset$, i.e. an evaluation process may refuse to indicate any subprograms in p as valuable. In PANGEA, this special case occurs when a classifier does not use any attributes from the dataset.

For instance, a decision tree inducer like C4.5 [150], when faced with poorly discriminating attributes, may produce a 'decision stump', i.e. a degenerate decision tree comprising single leaf that classifies all examples to the majority class.

Example 8.1. Consider the evaluation process in PANGEA illustrated in Fig. 7.5. The decision tree induced in that example used only attributes x_1 and x_4. The remaining attributes have been deemed not discriminative enough to be used by an inducer when constructing the decision tree. x_1 corresponds to an input variable x_1, i.e. to a single-instruction program (x_1). x_4 identifies in the evaluated program the subprogram $(x_2 + x + 1)$ (the right subtree of the program tree). Therefore, $P_C(p) = \{(x_1), (x_2 + x + 1)\}$. The valuable subprograms identified in this way in all candidate programs in population P can be then archived as described in the subsequent sections.   ■

8.2 Archiving subprograms

We are ultimately interested in reusing the subprograms 'harvested' from the candidate programs in a population. Valuable subprograms may emerge in any individual that undergoes evaluation, so it seems reasonable to gather them in a central repository. However, a few aspects have to be taken into account in that process.

code reuse

Firstly, subprograms originating in different programs in a population P cannot be considered equally valuable. The inducer will always use any available attributes to construct a possibly good classifier. In the case of the C4.5 inducer, to be included in the resulting classifier, an attribute needs to provide just slightly better conditional entropy than any other attributes (calculated locally, for a given branch of decision tree; [150]). In consequence, the usefulness of subprograms gathered in $P_C(p)$ may strongly vary across the programs $p \in P$.

Secondly, many inducers, including C4.5, are not limited in the number of attributes they select to construct a classifier. Which and how many attributes will be ultimately used by a decision tree depends not only on the above-mentioned entropy (*information gain*, to be more precise), but also on the heuristic strategy employed by an inducer. Consider two classifiers C_1 and C_2 induced from execution records of two different programs. C_1 may use five attributes, while C_2 fifty, yet they may still commit the same classification error (and even identically classify particular examples).

These premises motivate the design of a repository of subprograms, called hereafter an *archive*. The archive A is a prioritized queue of a fixed length,

archive

maintained throughout the entire evolutionary run. The priority is determined by a *utility measure* of a subprogram p' in program p:

sub-
program
utility

$$u(p') = \frac{1}{(1 + f_e(p))|P_C(p)|}, \tag{8.1}$$

where $f_e(p)$ is the error of the classifier induced from p's execution record (7.1). Within a single evaluation act, subprograms are thus rewarded with high utility if they facilitate building a good classifier (low $f_e(p)$) and if there are not too many subprograms (low $|P_C(p)|$). An ideal subprogram p' that alone ($P_C(p) = 1$) makes it possible to induce a perfect classifier ($f_e(p) = 0$) receives $u(p') = 1$. Conversely, if many subprograms are needed to build a classifier or/and the classifier has low predictive accuracy, $u(p')$ will gravitate to zero.

Each evaluation act produces a new batch of subprograms $P_C(p)$ and requires A to be accordingly updated. First, the subprograms from these two collections are temporarily merged into a single set $A' = A \cup P_C(p)$ and A is emptied. Then, we apply the utility-proportionate selection to A' (the counterpart of the fitness-proportional selection commonly used in EC). The selection results with a single subprogram p', which is added to A, unless there is already a subprogram $p'' \in A$ that is semantically equivalent to p', i.e. produces the same output for all considered tests (Sect. 5.3). In that latter case, only the smaller subprogram of p' and p'' remains in the archive and is granted with utility equal to $\max(u(p'), u(p''))$. This selection process repeats until A reaches its capacity or there are no more subprograms in A' that meet the above requirements. An evolutionary run starts with an empty archive, and its capacity is assumed to be low (in the order of dozens) in order to help proliferating the most promising building blocks.

This way of updating an archive provides for its semantic diversity, enables temporal variance, and promotes small subprograms, which is intended to reduce bloat when subprograms get reused in new candidate solutions. The quality of subprograms held there dynamically adapts to the current performance of programs in population: a subprogram that was considered valuable in the initial generations of a GP run may have no chance to survive in the archive once more useful subprograms emerge and become noticed by behavioral evaluation.

8.3 Reuse of subprograms

The objective of archiving subprograms is their reuse in new candidate solutions. For compliance with evolutionary framework of GP, it is natural to implement this functionality as a mutation-like search operator. Given a parent program tree p, the operator picks a random node in p and replaces

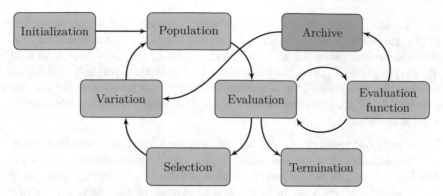

Fig. 8.1: The workflow of genetic programming synthesis system extended with an archive and behavioral code reuse. Compare to Fig. 1.1.

it (and the subtree rooted in it) with a subprogram drawn from the archive A proportionally to utility (8.1). Therefore, subprograms of higher utility are more likely to become components of other programs.

This search operator, though applied to a single parent program, can be considered as a form of crossover, because it fits the offspring with a piece of code that has been previously retrieved from another individual. Contrary to the conventional subtree-replacing mutation, no genuinely new subprograms and no instructions other than those currently present in the archive are being implanted in parent programs. In a longer run, this may cause some of the instructions to disappear from the population altogether. For this reason, we recommend to use this operators side-by-side with the conventional mutation or any other search operator that can provide population with a supply of fresh 'genetic material'.

8.4 Discussion

Figure 8.1 summarizes the workflow of behavioral code reuse proposed here by confronting it with the conventional GP workflow (Fig. 1.1). While the population is only a 'snapshot' of the current state of the search, an archive holds the less transient knowledge and serves as a centralized, long-term memory of search process. In this respect, it bears some similarity to more conventional archives used in EC that typically gather the best candidate solutions found so far, which makes particularly sense when in multi-objective setup (for instance, the NSGA-II algorithm referred several times in this book maintains an internal archive of nondominated solutions). However, in contrast to those archives that store entire candidate solutions, the archive considered here stores *fragments* of candidate solutions (subprograms).

In a broader, component-wise perspective, any stage of a GP algorithm might use the knowledge gathered in archive in order to advance search. An obvious alternative to the archive-based mutation proposed here is to feed the subprograms from archive directly into population, promoting them so to complete programs. This is admissible in type-less tree-based GP (or any closure   tree-based GP that ensures *closure* [79, 148]), where every subprogram is a valid program.

In the past GP research, several approaches have been proposed that maintain repositories of code pieces and engage code reuse. *Automatically defined functions* [78] can be seen as such repositories of subprograms, albeit local, associated with an individual, and devoid of any directed maintenance (i.e. the content of the repository is controlled only by evolution). Related research efforts quite often aimed at *knowledge transfer* between different problems in program synthesis and machine learning. *Run transferable libraries* [157] collect program fragments throughout a GP run and reuse them in separate evolutionary runs applied to other problems. Rosca and Ballard [153] used an analogous library within a single evolutionary run, with sophisticated mechanism for assessing subroutine utility, and entropy for deciding when a new subroutine should be created. Haynes [48] integrated a distributed search of genetic programming-based systems with 'collective memory', albeit only for redundancy detection. Other approaches involving some form of library include reuse of assemblies of parts within the same individual [58] and explicit expert-driven problem decomposition using layered learning [4]. Last but not least, the methods we proposed in [66, 68, 67, 102, 65] used GP archives for solving multi-class machine learning and pattern recognition problems, where we allowed multiple co-evolving populations delegated to particular decision classes reuse the code evolved in other populations.

The archive as proposed in this chapter differs from the work reviewed above mainly in the way the to-be-archived code pieces are determined. In the past approaches, that choice was typically based on the frequency of a subprogram occurrence in a population, or on subprogram's contribution to program's evaluation. In behavioral code reuse, such decisions are founded on test-wise and instruction-wise analysis of program behavior facilitated by execution records.

8.5 Summary

Subprogram archives presented in this chapter do not explicitly impose selection pressure on candidate programs in a population. They are thus inessential for behavioral evaluation, and should be rather considered as a complementary mechanism. On the other hand, as evidenced by our earlier results in [96] and in Chap. 10, they substantially improve the likelihood of solving program synthesis tasks. Also, as we show in Sect. 10.5, the behavior-aware selection of subprograms is essential for this mechanism to be effective.

This chapter concludes the part of this book that covers selected techniques designed to (more or less explicitly) broaden the evaluation bottleneck indicated in Sect. 2.1. In Chaps. 4–8, we arranged them with respect to increasing level of detail in which they peruse an execution record. This selection is by no means complete; arguably, there are other approaches not reviewed here that aimed at the same goal. In the next chapter, we attempt to gather the evaluation functions characteristic of such approaches under the common conceptual umbrella of search driver.

9

Search drivers

In this chapter, we provide a unified perspective on the methods presented in Chaps. 4–8, the key consequence of which is the concept of *search driver* detailed in Sect. 9.3.

9.1 Rationale for the unified perspective

In Chaps. 4–7, we presented several evaluation methods for characterizing candidate programs in GP. We summarize them in Table 9.1 and contrast with the conventional GP for reference.[1]

The approaches are founded on various formalisms. They rely on different parts of the execution record. Most of them evaluate programs in absolute terms, but some are relative and contextual, i.e. their assessments depend also on the other candidate solutions in a population. In the listed order, they are conceptually more and more sophisticated, and tend to elicit more information from an execution record. The last two listed methods can serve as subprogram providers (Sect. 8.1), i.e. can identify potentially valuable code pieces in evaluated programs.

These differences notwithstanding, all these alternatives to standard GP have been designed with a more or less explicit intention of broadening the evaluation bottleneck and acquiring alternative (or additional) *behavioral information* from candidate solutions (programs). We postulate that by this token they deserve a common conceptual umbrella, and from now on we refer to the evaluation functions they define as *search drivers*. search driver

---

[1] Even though SGP does not involve alternative evaluation functions, it allows the replacement of a problem-specific objective with a metric that, e.g., enables more efficient search operators (cf. [143]). By this token, it is also included in this table.

© Springer International Publishing Switzerland 2016
K. Krawiec, *Behavioral Program Synthesis with Genetic Programming*,
Studies in Computational Intelligence 618,
DOI: 10.1007/978-3-319-27565-9_9

Table 9.1: Summary of key properties of the approaches presented in Chaps. 4–7. GP: genetic programming equipped with conventional objective function. IFS: Implicit fitness sharing. DOC: discovery of underlying objectives by clustering. SGP: Semantic GP. PANGEA: Pattern-Guided Program Synthesis.

Method	Chap./ Sect.	Evaluation function	Part of execution record used	Objective	Source
GP	1	f_o (1.7)	Outcome vector	Yes	
IFS	4.2	f_{IFS} (4.4)	Outcome vector	No	[124]
Cosolvability	4.3	f_{cs} (4.7)	Outcome vector	No	[94]
DOC	4.4	f_{DOC}^j (4.11)	Outcome vector	No	[95]
SGP	5	d (5.3)	Program semantics	Yes	[135]
Trace consistency	6	f_{TC} (6.4)	Traces (equivalence)	Yes	[100]
PANGEA	7	f_e, f_c (7.1,7.2)	Traces (content)	Yes	[101]

As we argue in the following, search drivers are not required to objectively assess candidate solutions in the context of given problem – this is what the correctness predicate (1.1) is for. They are to *guide* search, meant as an iterative improvement process, by creating a gradient toward better performing solutions. The concept of search driver is thus a generalization of evaluation function.

To get a better grip on this new concept, let us recall first that objective function is not necessarily the best tool to guide a search (Sect. 2.2). On the other hand, none of the evaluation functions in Table 9.1 can be claimed *universally* best, as this would violate the *No Free Lunch Theorem* [193, 192]. Thus, rather than looking for an imaginary 'Holy Grail' of evaluation functions, we ask: can we characterize the minimal set of requirements which, when met by a given evaluation function, make it a useful (even marginally useful) tool to guide search? What is the minimal amount of feedback an evaluation function must provide so that we may talk about any guidance at all? Answering these questions in a rigorous manner will help us delineate search drivers.

No Free
Lunch
Theorem

9.2 Design rationale

Heuristic search algorithms alternate two actions: generation of new candidate solutions (variation) and their selection (cf. Fig. 1.1):

$$\ldots \xrightarrow{o} P_i \xrightarrow{sel} P' \xrightarrow{o} P_{i+1} \xrightarrow{sel} \ldots, \tag{9.1}$$

where $P_i \subset \mathcal{P}$ is the state of population in the ith iteration of search, and $P' \subset P_i$ is the sample (a multiset in general) of candidate solutions selected from P_i by the selection phase *sel*. The candidate solutions in P' form the basis for creating new candidate solutions with o, which may internally implement several specific search operators. The new candidate solutions populate P_{i+1}.

Formula (9.1) is a succinct phrasing of an Evolutionary Algorithm (EA) and virtually any iterative heuristic search algorithm. For instance, in a $\mu + \lambda$ evolutionary strategy [151], $|P_i| = \mu + \lambda$, and *sel* returns the μ best parents from P_i, which are then augmented by λ offspring by o, leading so to P_{i+1}. In local search, $P' = \{p\}$ and P_i is a sample of p's neighborhood. In an exhaustive local search, o generates the entire neighborhood P_{i+1} from the current candidate solution p, and *sel* selects the best candidate from it. In stochastic local search, o generates only a sample of neighborhood. Even random walk conforms to (9.1): in that case, o generates a random candidate solution, and *sel* is an identity function.[2]

Search operators in o may also utilize the evaluation outcomes, though in conventional EAs it is usually not the case, and in the following we will be agnostic about that aspect. For clarity, we also ignore for now initialization and termination (Fig. 1.1).

Evaluation is not explicitly present in (9.1). We assume that *sel* is 'informed' by some evaluation process. Conventionally, an evaluation function f is globally defined, static, and absolute, like the objective function f_o. The concept of search driver originates in observations that *these properties can be substantially relaxed*:

Observation 1: The evaluations $f(p)$ that guide *sel*'s decisions do not need to be independent across the candidate solutions $p \in P_i$. $f(p)$ may change subject to removal of any other $p' \in P_i$. This is the case when evaluation is contextual, e.g., for IFS, CS, and DOC in Chap. 4. In such cases, f's domain is effectively $\mathcal{P}^n$, where n is population size, and it returns an n-tuple of evaluations of its arguments.

Observation 2: The implementation of *sel* usually involves multiple invocations of a selection operator, each returning a single candidate solution from P_i picked from a relatively small sample $P'' \in P_i$ of candidates, while the characteristics of the remaining solutions in $P_i \setminus P''$ are irrelevant. At that particular moment, f does not even have to be defined outside of P''. For instance, tournament selection cares only about the evaluations of a few candidate solutions that engage in the tournament.

---

[2] Formula (9.1) is actually expressive enough to embrace *any* search algorithm, including exact algorithms like A^*, once we assume that P_i, rather than being a subset of $\mathcal{P}$, is a more general *state* of search process – a formal object that stores the knowledge about the search conducted to a given point, like the prioritized queue of states in A^*.

Observation 3:   For selection, it is sufficient to make *qualitative* comparisons on f for the candidate solutions in P_i (or in a sample $P'' \in P_i$ –
cf. Observation 2). The absolute values of f are irrelevant; it is *orders*
that matter. Tournament selection is again a good example here. Moreover, comparisons between candidate solutions do not necessarily need
to conform to the requirements of completer orders (in particular to
transitivity). Also, some candidate solutions can be arguably incomparable. Algorithms that conduct search only by qualitative comparisons
are sometimes referred to as *comparison-based algorithms* [179].

These characteristics form the minimal set of properties of search drivers
posited in the beginning of Sect. 9.1. They are deliberately very modest: a
typical evaluation function, like the objective function f_o, imposes a complete ordering on *all* candidate solutions in $\mathcal{P}$ and thus conforms Observations 1–3 by definition. Apart from these properties, a typical evaluation
function has usually other characteristics that are not essential for search
algorithm as defined above, and are in this sense redundant. The chasm
between the characteristics of common evaluation functions and the actual
needs of a selection operator in a search process (9.1) as expressed in Observations 1–3 motivate the concept of search driver presented in the next
section. For clarity, our further argument will primarily focus on individual candidate solutions and local search algorithms. However, we will also
present how this considerations generalize to population-based algorithms.

9.3 Definition

search
driver

A *search driver* for a solution space $\mathcal{P}$ is any non-constant function

$$h : \mathcal{P}^k \to \mathbb{O}^k, \tag{9.2}$$

where $k \geq 1$ is the *order* of the search driver driver, and $\mathbb{O}$ is a partially
ordered set with an outranking relation $\prec$ (precedence)[3]. When applied to a
k-tuple P of candidate solutions from $\mathcal{P}$, $h(p)$ returns a k-tuple of evaluations
(*scores*) from $\mathbb{O}$. We assume that the scores $o_i \in h(P)$ correspond one-to-
one to the arguments $p_i \in P$ and are invariant under permutations of the
arguments in P. Accordingly, we allow for abuse of notation and write $h(P)$
even when P is not a tuple but a set. In the following, $o_1 \prec o_2$ means that
o_2 is more desirable than o_1. One may alternatively say that $h(P)$ returns a
partially ordered set (*poset*) $(P, \prec)$.

behavioral
search
driver

Many of the search drivers considered in this book are behavioral. A behavioral search driver can be always implemented with help of execution record,
because an execution record is the complete account of program execution

---

[3] $\mathbb{O}$ should not be confused with $\mathcal{O}$, the type of program output (Sect. 1.1).

(Sect. 3.1). In this light, such search drivers could be redefined as mappings from a domain of execution records $\mathcal{E}$ to $\mathbb{O}$; however, to embrace also the non-behavioral search drivers, in the following we conform to the signature in (9.2).

A search drivers with completely ordered codomain $\mathbb{O}$ will be referred to as *complete search drivers*. The conventional scalar real- or integer-valued evaluation functions are examples of such drivers. The motivation for using partial orders is to allow search drivers to abstain from deeming some solutions better than others, which may be desirable when two solutions fundamentally differ in characteristics. As complete orders are special cases of partial orders, an evaluation function with the signature (9.2) that orders its arguments linearly is a search driver as well. On the other hand, search drivers generalize evaluation functions.

complete search driver

A search driver is *context-free* if the scores it assigns to its arguments are independent, i.e. it can be expressed by a one-argument function $f : \mathcal{P} \to \mathbb{O}$:

context-free search driver

$$h(p_1, \ldots, p_k) = (f(p_1), \ldots, f(p_k)).$$ (9.3)

Most of conventional evaluation functions f used in EC and GP are context-free search drivers in this sense. Note that for context-free search drivers, the order k is irrelevant. A search driver that is not context-free will be referred to as *contextual*. Note that the signature given in (9.2) is necessary to correctly define a contextual search driver: for instance, when defining $f_{\mathrm{IFS}}(p)$ (4.4) we silently allowed for an abuse of notation, as $f_{\mathrm{IFS}}(p)$ depends not only on p, but also on the other members of the population that p belongs to (the context).

contextual search driver

This definition of search driver follows the design rationale presented in the previous section. Firstly, h evaluates a *set* of candidate solutions (or more precisely a tuple) rather than individual solutions, and is thus contextual (Observation 1). A candidate solution p may receive different evaluations depending on the remaining elements of P, the argument of $h(P)$. Formally, $h(\ldots, p_i, \ldots) = (\ldots, o_i, \ldots)$ and $h(\ldots, p_j, \ldots) = (\ldots, o_j, \ldots)$ where $p_i = p_j$ does not imply that $o_i = o_j$, because the partial orders returned by h in these two applications can be in general completely unrelated. Secondly, the characteristics of the remaining candidate solutions in $\mathcal{P}$ do not have to be known (nor even computable) at the moment of applying h to a particular subset of them (Observation 2). Thirdly, the evaluations assigned to particular elements of P are not only allowed to be qualitative, but also only partially ordered (Observation 3).

Our rationale behind naming this formal object 'search driver' is twofold. The word 'search' signals that the class of problems we are primarily interested in here are search problems, even if they are disguised as optimization problems in GP. The 'driver' is to suggest that, other than creating *some* search gradient, a search driver is not promising to necessarily reach the search target (or detect the arrival at it) – in contrast to what the term 'objective function' suggests. In this sense, search drivers care more about *evolvability* than about reaching the ultimate goal of search.

The contextual evaluation functions discussed in Chap. 4 (IFS, CS, and DOC) can be phrased as complete search drivers, which we illustrate this with the following example.

Example 9.1. Refer to Example 4.1 of IFS evaluation and Table 4.1. The f_{IFS} evaluation function applied there to population $P = (p_1, p_2, p_3)$ produces $f_{\mathrm{IFS}}(p_1) = 2$, $f_{\mathrm{IFS}}(p_2) = {}^3\!/_2$, and $f_{\mathrm{IFS}}(p_3) = 1$ as the corresponding evaluations. An equivalent order-3 search driver h can be defined as

$$h(p_1, p_2, p_3) = (2, {}^3\!/_2, 1). \tag{9.4}$$

More generally,

$$h(p_1, p_2, p_3) = (a, b, c), \tag{9.5}$$

where $a, b, c \in \mathbb{O}$ could be abstract values ordered as follows: $c \prec b \prec a$.

Assume that we consider the differences on f_{IFS} for the pairs (p_1, p_2) and (p_2, p_3) too small to deem any of the compared candidate solutions better. In such a case, one could redefine the order in $\mathbb{O}$ so that so that only $c \prec a$ would hold.

Last but not least, the order of a search driver does not have to be bound with the size of population; an exemplary order-2 search driver for this problem could be defined as

$$h(p_1, p_2) = h(p_2, p_3) = h(p_1, p_3) = (a, b), \tag{9.6}$$

where $a \prec b$. ∎

9.4 Search drivers vs. selection operators

A vigilant reader might have noticed that search drivers can be directly used as selection operators. Indeed, $h(p_1, \ldots, p_k)$ is a k-tuple of values $(o_1, \ldots, o_k)$ from $\mathbb{O}$, and the maximal elements[4] in $(o_1, \ldots, o_k)$ are obvious candidates for being selected. Should $\mathbb{O}$ be only partially ordered, the selection between incomparable elements in $\mathbb{O}$ could be addressed by random drawing.

Though these similarities might suggest equivalence of search drivers and selection operators, we find it important to distinguish between these two

---

[4] In general, the maximal elements in the sense of partial orders. A partially ordered set may have arbitrary many maximal elements.

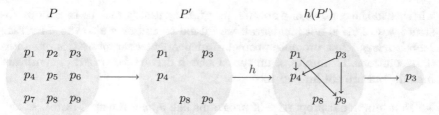

Fig. 9.1: The process of selection involving a single search driver. The selection operator draws a sample P' of candidate solutions from the population P. Search driver h is applied to P' and returns a partial ordering of the elements of P', in which it may deem some pairs of candidate solutions incomparable (e.g., p_1 and p_3). Finally, the selection operator makes a decision about selection based on h, yielding p_3.

concepts. A selection operator is a component of metaheuristic architecture (9.1) that implements the *entirety* of preferences concerning navigation in the search space as well as the desired characteristics of solutions. A single search driver, to the contrary, is an *elementary* source of information, reflecting only selected characteristics of candidate solutions. This distinction becomes particularly clear when using multiple search drivers simultaneously (Sect. 9.8). Thus, selection should be seen as a higher-level component that may engage one or more search drivers, applying them to accordingly prepared (typically drawn at random) samples of candidate solutions. We illustrate this principle in Fig. 9.1.

Search drivers differ from selection operators also in scope of application. A selection operator is typically applied to entire working population. The scope of search driver is determined by its order k, which we assume to be usually low compared to population size (recall that in Sect. 9.1 we set out to define the *minimal* feedback needed to effectively guide search).

Another difference is determinism. We define search drivers as deterministic functions, leaving the non-deterministic aspect of selection to a selection operator. In particular, we assume that it is the selection operator that is responsible for drawing a randomized sample of candidates, which are then passed to a search driver. Handling special cases with randomness (e.g., tie-breaking) is also delegated to selection operator in our conceptual model.

9.5 Universal search drivers

Remarkably, our definition of search driver in (9.2) does not refer to a program synthesis task. A search driver defines certain features of a solution, whether in the context of a specific program synthesis task or in a more abstract way. This is not incidental: we intend to embrace also the *universal search drivers* that promote problem-independent characteristics of candidate solutions. Examples of universal search drivers for program synthesis include, but are not limited to:

universal
search
driver

- Non-functional properties of programs like *program length* (size), *execution time*, *memory occupancy*, or *power consumption*.

- *Input sensitivity*. In cases where program input is a tuple of variables, it might be important to synthesize programs that take them all into account. There are two variants of this characteristic. A program can be said to be *syntactically sensitive* to all variables if it *fetches* them all (reads them in). A program is *operationally sensitive* if it can be shown that for every variable v_i there exists a combination of the remaining variables such that the output of a program changes when v_i is being changed. A syntactically sensitive program does not have to be operationally sensitive. The latter property is usually more desirable, but the former one is easier to verify.

- *Evolvability*. Given the iterative nature of search process, it is desirable to promote candidate solutions that can be subsequently modified to make further variation possible, and help reaching the optimal solution in a longer perspective. In tree-based GP, evolvability is hampered by, among others, bloat: standard search operators tend to extend deeper parts of programs (close to program leaves), while such changes often have no impact on the behavior of a program.

- *Smoothness*. In symbolic regression, programs are real-valued functions that are often required to be smooth, i.e. the output of a program should not change abruptly in response to modifications of the input.

Note that most of the above properties are behavioral, which is in tune with the leading motif of this book. Also, some of them are inherently qualitative; for instance, syntactic input sensitivity is basically a binary property. Some other characteristics are more quantitative but naturally deserve approximate comparisons. For instance, minor differences of program length are negligible in most applications: it really does not matter whether a symbolic regression model has e.g., 38 or 39 instructions. Another scenario that calls for tolerance is when the underlying measure is noisy. These examples show that defining search drivers in a qualitative manner is practical.

Example 9.2. An order-2 search driver that maps the exact program length onto a qualitative indicator can be defined as

$$h(p_1, p_2) = \begin{cases} (0,1), & if\,|p_1| > |p_2| + \beta \\ (1,0), & if\,|p_1| < |p_2| - \beta\,, \\ (0,0), & otherwise \end{cases} \tag{9.7}$$

where $|p|$ is the length of program p, β is the tolerance threshold, and $0 \prec 1$. This search driver renders any two programs that differ in length by less than β as indiscernible, and so imposes a complete order (pre-order to be precise) on its arguments. Note that the outranking relation defined by h is in this case intransitive.

Should it be more appropriate to use partial orders, the codomain could be extended to $\mathbb{O} = \{0, 1, \phi\}$, where $0 \prec 1$would be the only outranking in $\mathbb{O}$. Then, the third case in (9.7) would return (ϕ, ϕ), so that minor differences in program length would be interpreted as incomparability.   ∎

9.6 Problem-specific search drivers

A problem-specific search driver is a search driver that refers to the specification of program synthesis task. The k-tuples returned by such a driver depend not only on the k arguments (programs) but also on the *Correct* predicate of a task of consideration (Sect. 1.2). We assume the dependency on the latter to be implicit, unless otherwise stated.

The conventional objective function f_o may be trivially cast as the following complete, problem-specific, context-free search driver of an arbitrary order k:

$$h(p_1, \ldots, p_k) = (f_o(p_1), \ldots, f_o(p_k)). \qquad (9.8)$$

By (1.7), such a search driver depends on T, i.e. all tests that define a given program synthesis task. Alternatively, we may consider a search driver that counts program's failures in an arbitrary *subset* of tests $T' \subset T$. For instance, T' could have been determined by an underlying objective (4.10) in the DOC method presented in Sect. 4.4. A more sophisticated variant is an order-2 search driver based on the inclusion of passed tests

$$h(p_1, p_2) = \begin{cases} (0, 1), & if\ T(p_1) \subset T(p_2), \\ (\phi, \phi), & otherwise \end{cases} \qquad (9.9)$$

where we recall that $T(p) \subseteq T$ is the subset of tests passed by program p, and $\phi \in \mathbb{O}$ is the special incomparable value (cf. Example 9.2). Such h defines a partial order that implements the concept of *measure* on a set. An extreme case is a search driver that depends on programs' outcomes on a individual tests in T, i.e. dominance on tests

$$h(p_1, \ldots, p_k) = (g(p_i, t), \ldots, g(p_k, t)), \qquad (9.10)$$

where we recall that $g : \mathcal{P} \times \mathcal{T}$ is an interaction function (see 4.1).

The repertoire of problem-specific search drivers is by no means exhausted by the above examples. Virtually any evaluation function defined in past studies can be recast as a search driver in an analogous way. On the other hand, search drivers themselves represent a large class of functions. This helps convey that, in a sense, there is nothing special about the conventional test-counting objective function f_o: it is just one of many possible search drivers, but not necessarily the most effective one for a given program synthesis task (or a class of tasks). This observation leads to questions on quality of search drivers, which we address in the next section.

9.7 Quality of search drivers

We are ultimately interested in designing search drivers that perform well. However, what does 'perform' mean for a search driver? A search driver is just one component of iterative search algorithm, hidden inside the selection step in (9.1). The overall performance of the algorithm depends not only on the engaged search driver(s), but also on the selection operators, search operators, population initialization method, and possibly other components.

This suggests that it might be difficult, if not impossible, to investigate the the quality of a search driver in isolation from the remaining components of an iterative metaheuristic search algorithm[5]. Therefore, in this book we assume that the best gauge of a search driver's usefulness is its empirical performance on actual problem instances. In accordance with this, we introduce three categories of search drivers *for a given class of iterative search algorithms A* (9.1) and a set of problems (e.g., a suite of benchmarks). To this end, we define first the concept of random search driver.

random search driver

An order-k *random search driver* h is a search driver such that, for any given P and $(\ldots, o_i, \ldots, o_j, \ldots) = h(P)$ it holds

$$\Pr(o_i \prec o_j) = \Pr(o_j \prec o_i), \tag{9.11}$$

that is, it is equally likely that h orders o_i before o_j and that it orders them reversely. Note that in general $\Pr(o_i < o_j) \leq 1/2$ due to the potential presence of incomparability.

The random search driver serves as a reference point for defining *effective*, *deceptive*, and *neutral* search drivers:

effective search driver

• An order-k search driver is *effective* if it reduces the expected number of iterations of A compared to the number of iterations of A equipped with an order-k random search driver.

---

[5] In the process of writing this book, we have undertaken an attempt of designing a formal, algorithm-independent measure of search driver quality, based on the concordance of solution orderings provided by search drivers with the first hitting times of solutions. Unfortunately, the formalism required making many unrealistic assumptions, so we decided to not present it here.

- An order-k search driver is *deceptive* if it increases the number of iterations mentioned above.

- An order-k search driver is *neutral* if it does not affect the number of iterations in a statistically significant way.

We are obviously interested in effective search drivers, and with this book hope to pave the way for practical development and principled design thereof. From now on, by 'search driver' we will mean an effective search driver, unless otherwise stated.

We intentionally attribute deception to search drivers rather than to problems, even though the latter is prevailing in the literature. Recall that program synthesis tasks we consider here are inherently search problems that are only disguised as optimization problems (Sect. 1.5.3). A search problem cannot be deceptive, because all it defines is a set of candidate solutions (states) and the goal predicate (*Correct* in program synthesis). This the true, underlying information about the problem; no suggestion is being made that some of the candidate solutions are 'closer' to the search goal than others (whatever 'closer' would mean in this context). It is only a search driver (or an objective function in the more conventional setting) that can be deceptive in the above sense.

An *optimal order-k search driver* is a search driver h^* that maximizes, in the above sense, the performance of a given search algorithm on a given set of problems. This concept will serve us as a useful reference point in the following.

9.8 Employing multiple search drivers

Effective search drivers can be informally divided into *weak* and *strong* depending on how much they reduce the expected number of iterations required to reach the correct program. However, even weak drivers are assumed to perform significantly better than the random ones in the above sense.

Designing strong search drivers for program synthesis is difficult; if it was not, program synthesis would be a solved problem. In contrast, weak search drivers are by definition poor guides for a search process. Indeed, the studies we conducted earlier [97, 101] suggest that many search drivers do not work particularly well in isolation. However, given that search drivers may reflect different qualities of candidate solutions (see examples in Sects. 9.5 and 9.6), it is natural to consider using many of them.

There are many ways in which the judgments of multiple search drivers can be translated into decisions made by selection operator (or, in a wider context, the behavior of all algorithm components). We divide the techniques that facilitate usage of multiple search drivers into *sequential* and *parallel*.

sequential
usage of
search
drivers
The sequential methods allow different search drivers to be used at various stages of an iterative search process. In the simplest realization, the choice of search driver is explicitly controlled by the method. For instance, an analogous idea of alternating multiple evaluation functions (together with problem instances) at regular time intervals has been exploited to induce modularity in evolved neural networks [74, 75]. Interestingly, such proceed-

shaping
ing can be seen as yet another form of *shaping* which we touched upon in Sect. 4.2 [175].

Explicit control of the choice of search driver requires multiple design choices: in which order they should be used, for how many iterations, and whether they should take turns cyclically. Also, the transitions between particular search drivers suddenly change the 'rules of the game': adaptations acquired under one search driver may turn out to be maladaptations under subsequent search driver. Such 'catastrophic events' can be interesting when trying to reproduce some biological phenomena in silico (like in [74, 75]), but are not necessarily useful in program synthesis.

parallel
usage of
search
drivers
In this book, we argue for using multiple search drivers *in parallel*, primarily because this is consistent with our stance that no single search driver is a perfect means to control a search process, and thus no search driver should be favored. Another motivation is the above-mentioned problematic parameterization of sequential techniques. But even more importantly, parallel usage of search drivers has other appealing features discussed in below.

Partial independence. Assume n search drivers $h_i, i = 1, \ldots, n$. Let us denote by D_i the event that h_i orders two programs p_1 and p_2 discordantly with the optimal search driver h^*, i.e.

$$\Pr(o_1^i \prec^i o_2^i \wedge o_2^* \prec^* o_1^*). \tag{9.12}$$

Given the optimal nature of h^*, such discordance may result in an increase of the expected number of iterations. The probability that any pair of search drivers is simultaneously discordant with h^* is less or equal the probability that any of them is discordant. By the sum rule:

$$\Pr(D_i \cup D_j) = \Pr(D_i) + \Pr(D_j) - \Pr(D_i \cap D_j), \tag{9.13}$$

from which it follows that $\Pr(D_i \cap D_j) \leq \Pr(D_i)$, as by definition $\Pr(D_i \cup D_j) \geq \Pr(D_j)$. Simultaneous availability of h_i and h_j lowers thus the risk of mistakenly deeming one candidate solution less useful than another. The greater the number of search drivers, the less likely it becomes for them to be simultaneously discordant with the optimal search driver, and that likelihood is the lower the more independent are the search drivers in question.

inde-
pendent
search
drivers
In the extreme case of n *fully independent search drivers* , the probability of all of them being simultaneously discordant quickly vanishes with n:

$$\Pr\left(\bigcap_{i=1}^{n} D_i\right) = \prod_{i=1}^{n} \Pr(D_i). \tag{9.14}$$

On the other hand, the probability of all drivers being simultaneously *concordant* is also decreasing with n. However, if the search drivers in question are effective, they are likely to make more concordant decisions than discordant ones, i.e. $\Pr(o_1^i \prec^i o_2^i) < 1/2$, and the probability of at least half of n drivers to be concordant is greater than $1/2$. For the special case of $\Pr(D_i) = \Pr(D_j)$, $\forall i, j \in [1, n]$, that probability is determined by the cumulative distribution function of binomial distribution. Thus, if partially independent search drivers were to vote about a relation between a pair of candidate solutions, they are more likely to make the right decision than the wrong one.

Although designing a family of independent search drivers may be difficult, a certain degree of independence comes 'for free' for the search drivers summarized in Table 9.1 and examples given in Sects. 9.5 and 9.6, because they peruse different aspects of program behavior. Some methods promote independence on their own; for instance, the particular derived objectives built by DOC (Sect. 4.4) are based on disjoint subsets of tests, and may by that token be partially independent.

The above argument is analogous to the motivations for *committees* of classifiers in machine learning (a.k.a. *classifier ensembles*) [12]. ML committees are usually built to provide more robust predictions in the presence of noisy data. Just as multiple search drivers are less likely to simultaneously commit an error, so for the classifiers that vote about the output (decision class label or continuous signal) to be produced for a given example.

classifier ensembles

More specifically, partial ordering of candidate solutions performed by a search driver can be seen as an ordinal regression task (a partial one, to be more precise). A search driver is thus a special case of regression machine, and as such is characterized by certain *bias* and *variance* [38]. Bias represents a driver's inherent propensity toward certain realizations (models), while variance reflects the variability of a model's predictive accuracy. These quantities are inseparable, a characteristics known as *bias-variance tradeoff*: a highly biased predictor tends to have low variance and vice versa. However, by aggregating multiple low-bias, high-variance predictors, the variance can be reduced at no extra cost to bias. This observation is the key motivation for ensemble machines, and is naturally applicable also to search drivers.

bias-variance tradeoff

Diversity. By using several search drivers of different nature in parallel, we hope to provide for greater behavioral diversity in a population. Promoting behavioral diversity entails genotypic diversity, i.e. diversity of program code in the case of program synthesis. The importance of diversity maintenance has been demonstrated in population-based search and optimization techniques many times in the past, and was the major premise for designing methods like implicit fitness sharing (Sect. 4.2). We find diversity maintenance by means of behavioral search drivers particularly natural, as opposed to, e.g., niching techniques [118] and island models [191] that require parameter tuning.

multi-
modality

Multimodality. Program synthesis tasks are often multimodal, i.e. feature multiple optimal solutions, all of them conforming to the correctness predicate. A single-objective search process may be biased in tending to explore only selected basins of attraction of such optimal solutions. A search process that follows multiple objectives in parallel may be more open to explore many such basins. By the same token, multimodality is an argument against sequential usage of search drivers. Consider interlacing two search drivers along iterations of search process; if one of them happens to drive the search toward one optimum while another toward another optimum, search may cyclically oscillate between these optima.

Moderate computational overhead. Different search drivers may share algorithmic components needed to compute them. In such cases, calculating multiple search drivers rather than one does not necessarily incur massive overheads. For instance in TC (Chap. 6) and PANGEA (Chap. 7), recording of a program trace is a side effect of its execution and as such causes only moderate overhead.

9.9 Multiobjective selection with search drivers

In conventional GP, a scalar evaluation function serves as the basis for a straightforward selection operator (e.g., tournament selection). Simultaneous usage of several search drivers argued for in the previous section precludes direct application of such operators and requires special handling. In the following we discuss several alternative means to that end, most of which are only applicable when search drivers are complete, i.e. impose linear (pre)orders on candidate solutions.

Fig. 9.2 presents an example of a selection process involving two search drivers. As in the single-driver case (Fig. 9.1), the role of search drivers is to provide recommendation (a partial order of the considered sample of candidate solutions P'), while selection process is responsible for drawing P' and final selection of the 'winner'. Recommendations of particular drivers may be contradictory: for instance, p_1 and p_3 are incomparable according to h_1, while h_2 suggests that the former is worse than the latter.

The role of selection algorithm is to reconcile such discrepancies and appoint the best candidate solution within the considered sample P'. How to do

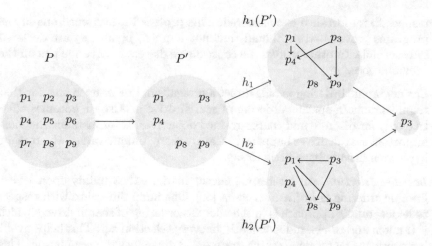

Fig. 9.2: The process of selection involving two search drivers h_1 and h_2. The selection operator has to reconcile the – partially contradicting – orderings provided by h_1 and h_2 to produce the final selection outcome, i.e. p_3. Consult Fig. 9.1 for an analogous single-driver example.

this appropriately and efficiently (particularly when the number of search drivers large) is in itself an interesting research question, which we do not address in this book. In the following, we discuss the possible aggregation methods for the case when all search drivers in question are complete, i.e. (pre)order the candidate solutions linearly.

Aggregation (scalarization). The arguably easiest way of handling multiple search drivers is to merge them into scalar evaluation, to be interpreted later by a single-objective selection method. If the codomains of search drivers in questions happen to be defined on metric scales, this may boil down to applying an averaging operator like arithmetic mean. Another alternative, geometric mean, is equivalent (up to an order) to the concept of *hypervolume* in multiobjective optimization. When search drivers range in different intervals and have different distributions, rank-based aggregation can be used.

hyper-volume

Scalar aggregation opens the possibility of using numerous conventional selection operators. On the other hand, aggregation incurs compensation (Sect. 2.2.2). Nevertheless, as we showed in [101], even a simple multiplicative aggregation of search drivers can offer substantial performance improvements.

Lexicographic ordering. A common multiobjective selection technique that avoids explicit aggregation is *lexicographic ordering*. This method expects the search drivers to be sorted with respect to decreasing importance. Given two candidate programs p_1 and p_2, they are first compared on the most important driver. If this comparison is conclusive, e.g., p_1 is strictly better than p_2 on that criterion, p_1 is selected. Otherwise, the next driver with

lexico-graphic ordering

respect to importance is considered. This process repeats until one of the programs proves better. Should that not happen, p_1 and p_2 are declared indiscernible. In other words, the consecutive drivers resolve the ties on the previous ones.

Lexicographic ordering avoids direct aggregation, but in exchange for that requires domain-specific ordering of search drivers. Also, it becomes effective only for discrete and coarse-grained objectives. If for instance the most important search driver happens to feature many unique values, the remaining search drivers have little to say.

Lexicase selection. An interesting recent method that builds upon lexicographic ordering is *lexicase selection* [50]. The main difference with respect to lexicographic approach is in the adopted ordering of search drivers which is random and drawn independently in every selection act. This helps avoiding overfocusing on some search drivers and diversifies the population. This straightforward and parameterless method proved very efficient in [50], where it was applied to tests, especially when the number of them was substantial. However, when applied to a moderate number of search drivers, its diversification capability and performance deteriorate [114].

lexicase
selection

In retrospect, lexicase selection can be seen as a test sampling technique. Such techniques typically draw a random sample of tests $T' \subset T$ in every generation of GP run, and use only the tests from T' for evaluation in that generation. Test sampling fosters diversity and improves performance, even when brought to extremes, i.e. drawing just a single test in each generation [43]. Lexicase selection 'individualizes' this process for particular selection acts.

Multiobjective selection. Multiobjective evolutionary algorithms offer several methods that avoid the pitfalls of aggregation while still eliciting useful information on search gradient. Usually, the underlying formalisms are dominance relation and Pareto ranking. Of the multiobjective selection methods, the Non-dominated Selection Genetic Algorithm (NSGA-II, [26]) is arguably most popular. NSGA-II employs a tournament selection on Pareto ranks to make selection choices. As a tie-breaker, it employs *crowding*, a measure that rewards the candidate solutions that feature less common scores on search drivers. The method is also elitist in selecting from the combined set of parents and offspring, rather than from parents alone. NSGA-II is the selection algorithm used in the experiment reported in Chap. 10. Many past works in GP proved the usefulness of multiobjective approach; see, e.g., [25], where an ad-hoc multiobjective algorithm was used for simultaneous promotion of diversity and reduction of program bloat.

9.10 Related concepts

There are several concepts in computational and artificial intelligence that bear some resemblance to that of a search driver.

In EC, the concept that arguably resembles search driver is *surrogate fitness*. Also known as *approximate fitness function* or *response surface* [71], a surrogate fitness function provides a computationally cheaper approximation of the original objective function. Surrogates are particularly helpful in domains where evaluation is computationally expensive, e.g., when it involves simulation. They usually rely on simplified models of the process being simulated, hence yet another alternative name: *surrogate models*. In continuous optimization, such models are typically implemented using low-order polynomials, Gaussian processes, or artificial neural networks. Occasionally, surrogate models have been also used in GP. For instance, in [51], Hildebrandt and Branke proposed a surrogate fitness for GP applied to job-shop scheduling problems. A metric was defined that reflected the behavioral similarity between programs, more specifically how the programs rank the jobs. Whenever an individual needed to be evaluated, that metric was used to locate its closest neighbor in a database of historical candidate solutions and neighbor's fitness was used as a surrogate.

> surrogate
> fitness
> approxi-
> mate
> fitness
> function
> response
> surface

Search drivers diverge from surrogate fitness in several respects. Firstly, surrogate functions are by definition meant to *approximate* the original objective function. Search drivers are, to the contrary, based primarily on the evidence that objective functions are not always the best means to navigate in a search space. Given the deficiencies discussed in Sect. 2.1 and the experimental evidence backing up the methods presented in Chaps. 4–7, why would one insist on approximating an objective function? Secondly, search drivers are primarily intended for search problems rather than optimization problems. This leaves more freedom in their design, which do not have to 'mimic' an objective function across the entire search space. Thirdly, in a program synthesis task, a search driver is not required to be consistent with a correctness predicate (1.5). In surrogate fitness, such consistency is essential.

Augmenting search with additional objectives is a part of the methodology proposed in [76] under the name of *multiobjectivization*. The additional objectives are introduced in that framework to make search more efficient, turn the original single-objective problem into a multiobjective one, and solve it using more or less off-the-shelf algorithms capable of handling multiobjective problems. An important assumption is that the extra objectives convey some *additional* problem-specific knowledge. In contrast, many of the search drivers discussed here essentially 'rephrase' the information that is conveyed – albeit subject to losses discussed in Chap. 2 – by the conventional objective function. The decomposition of scalar evaluation into multiple objectives in DOC (Sect. 4.4) is an example of such a proceeding. Also, with search drivers we put more emphasis on having many, even qualitative, information sources.

> multi-
> objec-
> tivization

The concept that is close to both multiobjectivization and search drivers is that of *helper objectives* [70], additional objectives used along with the original ('primary') objective in multiobjective setting. According to the cited

> helper
> objectives

work, helper objectives are meant to maintain diversity in the population, guide the search away from local optima, and help creation of good building blocks, meant as small components (*schemata*, to be more precise) that positively contribute to solution's evaluation [41]. The author argues that they should be 'in conflict' with the original objective function; for search drivers, we formalized a related concept of partial independence in Sect. 9.8. Helper objectives may change with time, i.e. only a subset of helper objectives is being used at any given time (dynamic helper objectives). The cited work also addresses the issue of size of Pareto front in multiobjective setting: once the number of candidate solutions with the same values of objectives exceeds the *niche count*, candidate solutions are randomly removed so that this constraint is not violated anymore. The approach was applied to job-shop scheduling and traveling salesperson problems, producing encouraging results.

Search drivers diverge from helper objectives in several respects. Helper objectives are meant to be used primarily with optimization problems – that is why the original objective function is always included as one of the objectives. Search drivers address search problems, in particular program synthesis, where achieving the sought solution is verified with a separate correctness predicate, so discarding the original objective function is acceptable. Also, designing helper objectives is similar to multiobjectivization in requiring substantial domain knowledge: for instance in application it to job-shop scheduling in [70], helper objectives reflected the flow-times of particular jobs. Later work also concerns job-shop scheduling [116] and confirms this deep immersion in problem domain. Search drivers, to the contrary, are more generic, and as we showed in Sect. 9.5, many of them are universal. Finally, search drivers are more general than helper objectives in being allowed to impose only qualitative and partial relationships between candidate solutions (see also the summary of search drivers' properties in Sect. 9.12).

In GP, several approaches have been proposed that aim at reducing the number of tests used for evaluation of candidate programs. In doing that, such methods effectively replace the original objective with an alternative evaluation function that can be sometimes likened to a search driver. The probably oldest contribution in this category is dynamic training subset selection [37]. In [115], selection of tests was applied to the task of software quality classification, in an attempt to reduce overfitting. In [43], this has been taken even further, i.e. single tests have been used. Several variants of this approach studied in [42] consistently reduced overfitting compared to standard GP.

heuristic
function

In AI, the concept that bears certain similarity to search driver is that of *heuristic function*. Heuristic functions in algorithms like A* bound the actual cost of reaching the search goal. Thy can be used to prioritize search and often thereby lead to performance improvements. They can be designed by relaxing the original problem, for instance, for an 8-puzzle problem, this can be achieved by assuming that any two neighboring tiles can be swapped, and counting the number of moves so defined. Interestingly, methods exist that generate heuristic functions automatically given problem formulation [156].

In reordering the visiting of candidate solutions, heuristic functions indeed resemble search drivers that may lead search in different directions than that of the objective function. However, heuristic functions explicitly rely on additional domain-specific knowledge that search drivers do without; examples include straight-line distance between cities in the famous Romanian roadmap example in [156] or the cost function for the 8-puzzle example mentioned above. Also, search driver is a more general formalism than heuristic function: unlike the latter, it is not guaranteed to bound the original objective function. As a consequence, algorithms that rely on search drivers cannot enjoy the 'comfort' of A*, which provably reaches an optimal solution (goal state) provided it exists in the searched tree. Providing analogous bounds in program synthesis is difficult due to the complex genotype-phenotype mapping (Sect. 1.4). In the typical tasks approached with the A*-like algorithms, the effects of search moves on the objective function are well understood, which facilitates designing efficient admissible heuristic functions. In program synthesis, a single move may change program's behavior almost arbitrarily.

In *reinforcement learning* (RL, [173]), the concepts of *intrinsic rewards* (or *intrinsic motivation*) [8] and internal rewards [169] bear distant similarity to search drivers. In nontrivial RL tasks, an agent often lacks external 'incentives' to explore and learn from an environment. Both methodologies mentioned above assume that, in the absence of such incentives, an agent would on its own 'induce' appropriate internal/intrinsic motivations and follow them, escaping thereby the detrimental state of temporal 'apathy'. In a sense, intrinsic rewards and intrinsic motivations can be seen as an agents's emanation of curiosity.

reinforcement learning

Last but not least, behavior search drivers share motivations with *novelty search* [110] (NS). In the spirit of *open-ended evolution*, often considered in artificial life community, novelty search discards search objectives altogether and rewards individuals for being behaviorally different from their peers in the current population and selected representatives of the search history. This turns out to work well on deceptive problems, especially on the problems where the mapping from genotypes to behaviors is strongly many-to-one, and the resulting behavioral space is relatively small. In [111], the authors applied novelty search to nontrivial GP benchmarks of maze navigation and Artificial Ant. The algorithm diverges from the traditional GP only in evaluation function. Prior to running evolution, an empty, unlimited-capacity archive A is created. Each evaluated individual has a low probability of being included in the archive. The evaluation of an individual in population P is defined as

novelty search open-ended evolution

$$f_{\mathrm{NS}}(p) = \frac{1}{k} \sum_{i=1}^{k} dist(p, \mu_i), \tag{9.15}$$

where $dist()$ is a behavioral distance measure, μ_i is the ith closest neighbor of p in $P \cup A$ with respect to $dist()$, and k defines the size of the neighborhood.

They key component of NS is the behavioral distance measure $dist()$. In [111], it was based on the final location in the maze problem and the time-wise distribution of food collections in the Artificial Ant problem. The authors of [121], the probably first attempt to apply novelty search to generic GP tasks, based $dist()$ on behavioral descriptors closely resembling outcome vectors defined in this book (2.3). The ith vector element is 1 if the individual belongs to an assumed (low) percentile of programs that commit the smallest error on a test; otherwise it is set to 0. Experimental assessment of this configuration on three benchmarks did not yield particularly conclusive results. Other works related to NS include, among others, [196] and [195], where diversification was promoted by maximizing correlation distance between time-wise behavior of GP programs representing trading strategies.

In general, we anticipate NS to struggle when faced with more demanding program synthesis tasks because of the size of behavioral space. Even in the simplest case when tests can be only passed or failed, the number of all possible behavioral descriptors grows exponentially with the number of tests, and becomes staggering even for the small benchmarks typically considered in GP (e.g., 2^{64} for the humble 6-bit multiplexer; see Chap. 10). One may doubt if simply enticing an evolving population to spread across such a space is sufficient to locate the target solution.

Nevertheless, NS shares motivations with behavioral program synthesis: Lehman and Stanley state that "The problem is not in the search algorithm itself but in how the search is guided" [111, p. 842], which strongly resonates with the arguments in this book. NS can be seen as the opposite extreme to conventional search driven by the conventional objective function. Search drivers sit in between these two extremes: they impose different search gradients than the conventional objective function, yet, contrary to NS, those gradients are not entirely detached from it.

9.11 Efficiency

Abandoning search objectives in favor of search drivers has measurable consequences for more technical aspects of implementation. Below we discuss such implications for the arguably most important technical aspect of program synthesis – computation cost. In a typical GP run, the lion's share of computation is spent on running programs, i.e. applying them to tests. No wonder the number of evaluated programs is the most common unit of computational expense in GP.

In Sect. 9.8, we argued for using several search drivers in parallel, in a multiobjective setting. As some search drivers incur substantial overheads (e.g., classifier induction in PANGEA), this may seem computationally prohibitive. However, the execution record stores the complete account of program execution for a considered set of tests. Once it has been computed for a given program (which is not particularly expensive as demonstrated in Sect. 3.2) most search drivers can be calculated from it at a moderate cost, because the data they require (outcome vectors, program semantics, program traces – see Table 9.1) can be immediately retrieved from the record (Fig. 3.2). Therefore, if the number of tests is large or/and program execution is in itself expensive, the total overhead resulting from using multiple search drivers may become negligible.

Moreover, in some scenarios search drivers allow for substantial *reduction* of computational expense. For instance, the convergence of execution traces in Chap. 6 can be used to reduce the time spent on program execution: if two traces merge (which the method has to detect anyway), i.e. program execution leads to the same execution state for two tests, then from that state on only one execution has to be conducted, as the other must proceed in exactly the same way.

Opportunities for potential savings wait to be uncovered not only in the internals of program execution, but also on the higher abstraction levels. Consider an order-2 search driver h that completely orders the programs according to the number of tests they pass. This search driver can be trivially expressed using the objective function f_o: $h : \mathcal{P}^2 \to \{0,1\}^2$, where $0 \prec 1$ and

$$h(p_1, p_2) = \begin{cases} (0,1), & \text{if } f_o(p_1) > f_o(p_2) \\ (1,0), & \text{if } f_o(p_1) < f_o(p_2) \\ (0,0), & otherwise \end{cases} \quad (9.16)$$

(recall that f_o is minimized). This formulation assumes that $f_o(p_1)$ and $f_o(p_2)$ are calculated independently, which for a set of tests T requires the total of $2|T|$ executions of p_1 and p_2. Now consider an algorithm that iterates over T and applies p_1 and p_2 to a given $t \in T$ simultaneously. Let $i \in [1, |T|]$ be the number (index) of the currently processed test, and $f_o^{(i)}(p)$ the number of tests failed so far. Note that as soon as the following condition starts to hold

$$|f_o^{(i)}(p_1) - f_o^{(i)}(p_2)| > |T| - i, \quad (9.17)$$

the loop over i can be terminated, because the outcomes for the remaining $|T|-i$ tests cannot compensate the already gathered evidence in favor of one of the programs. In such a scenario, the total number of program executions amounts to $2i$, and may be substantially smaller than $2|T|$ above. Though the actual benefits resulting from enhancements like this one are domain- and problem-dependent, even a minor reduction of the number of program executions may be beneficial in challenging program synthesis tasks.

9.12 Summary

This chapter described a preliminary attempt to crystallize the concept of search driver, a generalization of evaluation function intended to meet the needs of metaheuristic search algorithms and program synthesis in particular. Further effort is clearly required to get a better grip on it and possibly lead to principled design of search drivers. Nevertheless, it should be clear already at this point that search drivers exhibit common characteristics (cf. Observations formulated in Sect. 9.2):

1. Search drivers are contextual. The role of a search driver is to provide gradient within a relatively small set of candidate solutions. A search driver is not required to provide such a gradient globally.

2. Search drivers provide qualitative, ordinal feedback. Absolute values are irrelevant. What matters is the (partial or complete) order of candidate solutions.

3. Search drivers do not have to relate to the original objective function, and in particular do not have to correlate with it. Preferably, they should approximate the optimal search driver.

4. Search drivers do not have to be consistent in the sense of (1.5), i.e. to indicate the optimality of candidate solutions by achieving extreme values at them (nor in any other way). This functionality is delegated to a correctness predicate.

5. Search drivers may depend on the entire state of the search process meant as the working population of candidate solutions (or even state in a broader sense, for instance including candidate solutions visited in previous iterations).

6. As a consequence of (5), search drivers may be non-stationary, i.e. order the same subset of candidate solutions differently in particular iterations of a search loop.

7. Search drivers can be weak, i.e. order relatively many candidate solutions differently than an optimal search driver. Using many weak drivers in parallel can make the search process effective by providing sufficiently strong search gradient *and* diversity.

10

Experimental assessment of search drivers

This chapter presents the results of a comparative experiment involving various combinations of search drivers. Our goal, beyond demonstrating the strength of behavioral evaluation, is to answer the following questions:

1. What is the impact of particular search drivers when used in combination with other search drivers? The experiment addressing this question is reported in Sect. 10.3.

2. How useful is the mechanism of archiving useful subprograms presented in Chap. 8? (Sect. 10.4).

Our goal is to gain better understanding of behavioral search drivers and interactions between them, rather than engaging into an 'up-the-wall' game [17] with other genres of GP as competitors. Compared to our previous results reported in [96], here we consider more combinations of search drivers, and combine the drivers of PANGEA, trace consistency approach, and IFS. We also simplify archive maintenance, verify the importance of subprogram selection (Sect. 10.5) and investigate other aspects.

10.1 Scope

We consider only selected search drivers to avoid combinatorial explosion when considering their combinations. The experiment involves universal search drivers and those defined for trace consistency analysis (Chap. 6), pattern-guided GP (Chap. 7), and IFS (Chap. 4). We consider six search drivers in total, denoting them with the following mnemonics:

- F (f_o), the conventional objective function, i.e. the number of failed tests (1.7),

© Springer International Publishing Switzerland 2016
K. Krawiec, *Behavioral Program Synthesis with Genetic Programming*,
Studies in Computational Intelligence 618,
DOI: 10.1007/978-3-319-27565-9_10

- S (f_s), program length (size), i.e. the number of nodes in a GP tree,

- L (f_e), the error of the classifier induced from execution traces (7.1),

- C (f_c), the complexity of the classifier induced from execution traces (7.2),

- E (f_{TC}), the two-way entropy induced from execution traces (6.4),

- I (f_{IFS}), implicit fitness sharing (4.4).

All these drivers are complete, i.e. order linearly the compared programs. Among them, program size S is the only non-behavioral search driver and the only universal search driver; all the remaining ones are behavioral and problem-specific. f_{IFS} (I) is the only contextual search driver – all the remaining ones are context-free.

There are $2^6 - 1 = 63$ nonempty combinations of these drivers. However, IFS is the only contextual search driver and will be considered separately in Sect. 10.6, which halves for now the number of combinations to 31. We do not run experiments with individual search drivers as we anticipate them to perform poorly; for instance, one cannot expect to synthesize a correct program by driving synthesis with program size alone. Also, in Sect. 9.8 we brought several arguments in favor of using multiple search drivers in parallel. Therefore, we conduct single-objective synthesis only for F.

Additionally, we assume that every considered combination of drivers must include F. Also, C requires an ML classifier to be induced (Sect. 7.2.2), and the presence of a classifier implies availability of L, so C will be used only when L is present in a given combination. These constraints leave us with 11 combinations:

- Five purely behavioral combinations, i.e. not including program size: FL, FLC, FLE, FE, FLCE,

- Six combinations with program size: FS, FLS, FLSC, FLSE, FSE, FLSCE.

For each considered combination of drivers we run a tree-based generational GP algorithm with parameter settings summarized in Table 10.1. NSGA-II algorithm with tournament of size 7 is used for selection ([26]; see Sect. 9.9). As its computational complexity is quadratic in population size, we allow only 300 programs in the population. Also, in NSGA-II the parents and offspring are Pareto-ranked together, which makes the selection process highly elitist. The risk of losing a good candidate solution is thus small, so we permit quite intense exploration, setting the probability of mutation to 0.2, higher than the default 0.1 typically used in GP studies. The programs in initial population are relatively small to enable gradual 'complexification' of solutions.

Table 10.1: The common parameter setting for all considered methods.

Parameter	Setting
Population size	300
Population initialization	Ramped half-and-half, max. tree height 3
Selection operator	Tournament of size 7
Node selection operator	Uniform-depth selector
Mutation operator	Replacement with random subtree of height 3, prob. 0.2
Crossover operator	Subtree-swapping crossover, prob. 0.8
Offspring acceptance	Tree height ≤ 9
Maximum generations	300

The original NSGA-II algorithm [26] resolves ties between solutions in the same tier of Pareto ranking by resorting to *crowding*, a measure based on distance in objective space. This approach is rational in conventional multiobjective problems, where the goal is to obtain a representative, evenly distributed approximation of the true, unknown Pareto front, which may be then subject to choice of a final solution. Here, uniformity of that approximation is less important, because search drivers are only meant to guide search and their values become irrelevant once a solution is found (program size being the only exception). It is however desirable to promote *unique combinations* of objectives' values to promote diversity in population. Therefore, we define crowding for a given candidate solution p as the number of candidate solutions (in the same rank of Pareto front) that have the same evaluations as p on all objectives. Lower values of crowding are preferred when programs tie on Pareto ranks. In contrast to [26], we do not promote the solutions located at the extremes of Pareto fronts, as this may lead to pathologies: in a preliminary experiment involving S, degenerate one-instruction programs tended to win a very high fraction of tournaments, causing search to stall.

To lower the risk of bloat, we abandon the conventional methods for selecting a node (instruction) to be modified by search operators in a program tree. Normally, nodes are selected at random, with equal probability, and with optional special handling of tree leaves, like in the Koza-style node selector [79]. Rather than that, we employ *uniform depth node selector*, which, given a tree of height d, draws a random number d' from $[0, d]$ and returns a randomly chosen node at depth d'. In this way, the probability of selection does not grow exponentially with node depth, as it is the case in conventional methods. Mutations and crossovers at shallow depths are more likely, and tendency for bloating is reduced. Preliminary experiments showed that this is beneficial for all configurations, including standard GP.

However, even with the above node selector, the configurations devoid of S tend to produce large programs. We include thus a feasibility check: if the height of an offspring program tree exceeds 9, it is discarded and a new offspring is generated (with any search operator, according to the probability distribution in Table 10.1). As all instructions considered here are binary, this limits program size to $2^{9+1} - 1 = 1023$. Programs approaching this limit are very improbable, as evolving a perfectly balanced binary tree is unlikely.

In the configurations that involve L, we apply the REPTree algorithm [47] as the ML algorithm for inducing decision trees from execution records (Chap. 7). REPTree is much faster than traditional decision tree algorithms like C4.5 [150] and can induce both classification and regression trees. By default, it post-prunes (simplifies) the induced trees, but we disable this option and make it always split a decision node if that leads to an improvement of intra-node class consistency. In this way, search driver C (the size of decision tree) reflects better the complexity of mapping from execution traces to desired outputs, which is its very essence (Sect. 7.2.3)

A run terminates when a correct program is found, i.e. when $f_o(p) = 0$ (cf. (1.5)), or when 300 generations have elapsed.

We define two single-objective baseline setups:

- F, the conventional GP with f_o as the only objective,

- F+, as F, however with population of size 1 000 evolving for up to 90 generations.

The motivation for F+ is 300 candidate solutions can be considered too few for conventional GP, so F+ uses a larger population while maintaining the same overall computational budget of 90 000 evaluations per run.[1]

Both single- and multi-objective configurations employ tournament of size 7 to select the parents, where the former perform selection on the conventional evaluation function f_o, and the latter on the ranks of the Pareto ranking built in a multiobjective space spanning search drivers (and with crowding to resolve ties). This is the only difference between F and F+ and the remaining configurations.

The algorithms were implemented using software libraries prepared in Scala[2] and available online. The experimental configurations and detailed results can also be downloaded [82]. Scala is a succinct object-oriented and functional programming language that offers traits, mixins, lazy evaluation, built-in support for parallelism, advanced reflection mechanisms, and many other modern features. It is fully interoperable with Java and runs on its virtual machine.

---

[1] Technically, F and F+ could be alternatively implemented using NSGA-II with f_o as the only objective. However, this may result in undesired side effects; for instance, all solutions with the same rank have by definition the same value of crowding when evaluated on one objective.

[2] http://www.scala-lang.org/

Table 10.2: The program synthesis tasks used in the computational experiment.

Domain	Instruction set	Problem	Input variables	Tests	Unique semantics
		Cmp6	6	64	2^{64}
		Cmp8	8	256	2^{256}
		Maj6	6	64	2^{64}
Boolean	and, nand, or, nor	Maj8	8	256	2^{256}
		Mux6	6	64	2^{64}
		Par6	6	64	2^{64}
		Par8	8	256	2^{256}
Categorical	a_i	D_i	3	27	3^{27}
		M_i	3	15	3^{15}

10.2 Program synthesis tasks

Table 10.2 presents the 17 benchmark program synthesis tasks used in the experiment, 7 of which come from the Boolean domain and 10 from a categorical domain. The table lists the instruction sets used for each domain, and the number of variables, tests, and cardinality of search space for every problem. Note that there are no constants in instruction sets.

The targets for particular Boolean problems are defined as follows. For a v-bit comparator CMPv, a program is required to return *true* if the $\frac{v}{2}$ least significant input bits encode a number that is smaller than the number represented by the $\frac{v}{2}$ most significant bits. For majority MAJv problems, *true* should be returned if more that half of the input variables are *true*. For the multiplexer MULv, the state of the addressed input should be returned (6-bit multiplexer uses two inputs to address the remaining four inputs). In the parity PARv problems, *true* should be returned only if the number of inputs set to *true* is odd.

The categorical problems come from Spector et al.'s work on evolving algebraic terms [171] and dwell in a nominal ternary domain: the admissible values of program inputs and outputs are $\{0, 1, 2\}$. The peculiarity of these problems consists of using only one binary instruction in each benchmark, denoted by a_i in in Table 10.2. a_i is defined as follows for the considered five algebras:

a_1	0 1 2
0	2 1 2
1	1 0 0
2	0 0 1

a_2	0 1 2
0	2 0 2
1	1 0 2
2	1 2 1

a_3	0 1 2
0	1 0 1
1	1 2 0
2	0 0 0

a_4	0 1 2
0	1 0 1
1	0 2 0
2	0 1 0

a_5	0 1 2
0	1 0 2
1	1 2 0
2	0 1 0

For each algebra, we consider two tasks (of four discussed in [171]). In *discriminator term* tasks (D_i in Table 10.2), the goal is to synthesize an expression that accepts three inputs x, y, z and is equivalent to

$$t(x, y, z) = \begin{cases} x & \text{if } x \neq y \\ z & \text{if } x = y \end{cases}. \tag{10.1}$$

Given three inputs and ternary domain, this gives rise to $3^3 = 27$ tests. The second task defined for each of algebras (M_i in Table 10.2) consists in evolving a ternary *Mal'cev term* that satisfies

$$m(x, x, y) = m(y, x, x) = y. \tag{10.2}$$

When x, y, and z are all distinct, the desired output is not determined by this condition. There are thus only 15 tests in Mal'cev tasks, the lowest of all considered benchmarks.

The readers familiar with GP have surely noticed the absence of regression problems in our benchmark suite. We disregard them for a few reasons. Firstly, as stated in Sect. 1.2, program synthesis in its core formulation is a search problem, and not an optimization problem that symbolic regression inherently belongs to. Program correctness becomes elusive for continuous outputs. In practice, it is implemented by imposing a threshold on program error ($f_o(p) < \theta$), but it is hard to justify the setting of this parameter, unless it is given as a part of problem statement (which is not the case in benchmarks normally considered in GP). It is common to use very small θ, just to account for roundoff errors and render correct only the programs which, mathematically speaking, implement the target perfectly. However, aiming at perfect regression is questionable given that real-world data is always noisy, and resembles more interpolation than regression. To evade these dilemmas, we disregard symbolic regression problems in this study.

10.3 Combinations of search drivers

The 11 multiobjective configurations and the two single-objective ones, together with the considered 17 benchmarks, lead to the total of $13 \times 17 = 221$ configuration-benchmark pairs. To achieve statistical significance, for each of them we ran 50 evolutionary runs with a different seed of random number generator.

In consistency with our emphasis on producing correct rather than approximate programs, our primary performance measure of interest is *success rate*, i.e. the percentage of runs that produce a correct program, i.e. an unbiased estimate of the probability of synthesizing a correct program. Table 10.3 presents this performance measure for particular configurations and benchmarks. The conventional single-objective GP (configurations F and F+) tends to solve only the easiest benchmarks, even when working with a larger population (F+). The multiobjective methods based on two or more

Table 10.3: Success rates for particular combinations of search drivers.

	CMP6	CMP8	DISC1	DISC2	DISC3	DISC4	DISC5	MAJ6	MAJ8	MAL1	MAL2	MAL3	MAL4	MAL5	MUX6	PAR5	PAR6
FE	0.32	0.02	0.02	0.04	0.32	0.00	0.10	0.22	0.00	0.40	0.34	0.20	0.16	0.82	0.94	0.00	0.00
FL	0.34	0.00	0.00	0.00	0.08	0.00	0.02	0.22	0.00	0.08	0.14	0.24	0.04	0.68	0.98	0.02	0.00
FLC	0.62	0.12	0.02	0.02	0.24	0.00	0.04	0.46	0.00	0.28	0.16	0.38	0.20	0.74	0.78	0.06	0.00
FLCE	0.52	0.04	0.10	0.20	0.38	0.16	0.26	0.38	0.00	0.56	0.48	0.32	0.50	0.66	0.84	0.04	0.00
FLE	0.36	0.00	0.02	0.10	0.26	0.04	0.12	0.32	0.00	0.48	0.42	0.28	0.36	0.78	0.98	0.00	0.00
FLS	0.00	0.00	0.00	0.00	0.00	0.00	0.00	0.02	0.00	0.06	0.02	0.00	0.02	0.64	0.10	0.00	0.00
FLSC	0.20	0.00	0.00	0.00	0.00	0.00	0.00	0.14	0.00	0.14	0.12	0.12	0.24	0.86	0.46	0.00	0.00
FLSCE	0.12	0.00	0.04	0.02	0.14	0.02	0.00	0.04	0.00	0.34	0.30	0.10	0.42	0.88	0.30	0.00	0.00
FLSE	0.04	0.00	0.00	0.02	0.00	0.00	0.02	0.00	0.00	0.42	0.22	0.08	0.32	0.74	0.10	0.00	0.00
FS	0.00	0.00	0.00	0.00	0.00	0.00	0.00	0.00	0.00	0.00	0.00	0.00	0.00	0.58	0.04	0.00	0.00
FSE	0.00	0.00	0.00	0.00	0.00	0.00	0.00	0.00	0.00	0.18	0.16	0.00	0.08	0.90	0.14	0.00	0.00
F	0.36	0.02	0.00	0.00	0.00	0.00	0.00	0.56	0.00	0.14	0.06	0.26	0.00	0.84	0.92	0.04	0.00
F+	0.70	0.16	0.00	0.00	0.02	0.00	0.00	0.86	0.00	0.20	0.12	0.58	0.00	0.94	1.00	0.06	0.00

search drivers fare much better. All benchmarks, except for MAJ8 and PAR6, are solved at least once by some combinations of search drivers.

The performance on individual benchmarks does not tell much about a configuration; what ultimately matters are the relations between the likelihoods of synthesizing a correct program for any problem. In order to take into account that some benchmarks are inherently easier and some more difficult, we summarize Table 10.3 by ranking methods on every benchmark independently, and then averaging the ranks for each method:

FLCE	FLE	FLC	FE	F+	FLSCE	F	FL	FLSE	FLSC	FSE	FLS	FS
3.26	4.47	4.94	5.47	5.59	6.35	7.32	7.47	8	8.32	8.94	10.2	10.7

The ranking reveals that including program size S as a search driver is detrimental in all cases. Except for FLSCE, all configurations that involve S rank lower than F (conventional GP), and FLSCE ranks worse than F+. Preferring the smaller programs leads to elimination of the larger ones, and it is the latter programs that are evolvable, because they lend themselves to modifications that are harmless and prospectively beneficial (for instance when applied to *introns*, i.e. code pieces that do not affect program output). Small programs, to the contrary, tend to be more brittle: most modifications applied to them are usually detrimental.

Individual search drivers clearly complement each other. When used only with the conventional evaluation function F (FE, FL, FS), they perform rather poorly. When combined in groups of three (FLC, FLE, FLS, FSE), they fare better. When excluding S, the average rank of triples of search drivers amounts to 4.705 and the average rank for pairs of drivers is 6.47. It is the most numerous combinations that take the lead. Of them, FLCE fares the best, achieving success rate of 32 percent on average.

It is interesting to see that greater numbers of search drivers do not harm the success rate, even though multiobjective selection is known to become

Table 10.4: Average ranks of configurations that include and exclude particular search drivers.

Search driver	L	C	E	S
Configurations including the driver	6.63	5.72	6.08	8.75
Configurations excluding the driver	7.60	7.57	7.79	5.50

less efficient with the growing number of objectives. This confirms the hypothesis formulated in Sect. 9.8 that using multiple uncorrelated search drivers can make a search process more effective.

For statistical significance we conduct Friedman's test for multiple achievements of multiple subjects [72] which, as opposed to ANOVA, does not require the distributions of variables to be normal (which we cannot assume here). The p-value for the data in Table 10.3 is $\ll 0.001$, which strongly indicates that at least one method performs significantly different from the remaining ones. The post-hoc analysis using the symmetry test [57] indicates that the leading configuration FLCE is significantly better than FLSCE and all configurations that follow in the above ranking.

To assess the contributions of each driver, in Table 10.4 we confront the average ranks of configurations that include a given search driver with those that do not. The only driver that deteriorates performance is S. All the remaining drivers, when added to a combination of remaining drivers, improve its rank in expectation. The impact of C seems to be the greatest, improving rank by 1.85 on average.

10.4 Configurations with subprogram archives

The rationale for archives of subprograms (Chap. 8) is that by inspecting execution traces we obtain not only quantitative feedback in form of search drivers, but also subprograms that are relevant for the task being solved. Here, the attributes selected by the REPTree-induced decision tree (corresponding to instructions in an execution record and, indirectly, to a program subtree) indicate the relevance of particular subprograms in an evaluated program. The subprograms are gathered in an archive and reused by a customized search operator.

In this experiment, we simplify the workflow originally used in [96] and described in Chap. 8. As in that work, after every generation we gather all indicated subprograms in a working set A' and eliminate semantic duplicates: for any group $P_{eq} \subseteq A'$ of semantically equivalent subprograms, we preserve the smallest one, set its utility to the maximum of the utilities of subprograms in P_{eq}, and discard the remaining elements in P_{eq}. However, rather than fixing the archive's capacity, we discard half of A' in every

Table 10.5: Success rates for particular combinations of search drivers (configurations with subprogram archive).

	CMP6	CMP8	DISC1	DISC2	DISC3	DISC4	DISC5	MAJ6	MAJ8	MAL1	MAL2	MAL3	MAL4	MAL5	MUX6	PAR5	PAR6
FLA	0.98	0.66	0.16	0.40	0.84	0.04	0.32	0.92	0.00	0.80	0.80	0.98	0.78	0.96	1.00	0.68	0.22
FLCA	1.00	0.76	0.22	0.58	0.96	0.06	0.40	0.88	0.02	0.80	0.88	0.98	0.94	1.00	1.00	0.38	0.08
FLCEA	1.00	0.70	0.54	0.68	0.94	0.40	0.70	0.90	0.02	0.84	1.00	0.98	0.98	1.00	1.00	0.40	0.12
FLEA	1.00	0.64	0.32	0.44	0.92	0.28	0.54	0.96	0.00	0.98	1.00	1.00	1.00	1.00	1.00	0.72	0.30
FLSA	0.88	0.00	0.00	0.02	0.32	0.00	0.00	0.26	0.00	0.56	0.52	0.66	0.54	0.92	1.00	0.08	0.00
FLSCA	1.00	0.30	0.10	0.30	0.96	0.02	0.12	0.94	0.00	0.88	1.00	0.94	1.00	0.98	1.00	0.62	0.10
FLSCEA	1.00	0.18	0.28	0.74	0.88	0.10	0.34	1.00	0.00	0.98	1.00	0.94	1.00	1.00	1.00	0.20	0.00
FLSEA	0.84	0.00	0.14	0.46	0.88	0.10	0.18	0.66	0.00	0.94	1.00	0.96	1.00	0.96	1.00	0.12	0.02

generation, removing so the subprograms with utility below the median. Should this result in archive size dropping below 50, we keep the 50 subprograms with highest utility. The archive is emptied and repopulated with subprograms in every generation.

We also simplify the definition of subprogram utility. Rather than (8.1), we define it for a subprogram p' in a program p as

$$u(p') = \frac{1 - f_e(p)/|T|}{|P_C(p)|}, \qquad (10.3)$$

where $f_e(p)$ is the error of the classifier induced from p's execution record (7.1), $P_C(p) > 1$ is the number of subprograms/attributes used by that classifier, and $|T|$ is the number of tests. The 'reward' (classification accuracy) is evenly shared between the derived subprograms, while in (8.1) it depended nonlinearly on f_e.

To reuse code pieces from the archive, we use *archive-based mutation* described in Sect. 8.3. It works as subtree-replacing mutation, the only difference being that the program implanted in the parent is fetched from the archive rather than randomly generated. The new operator is granted half of the engagement probability of crossover (0.8, Table 10.1), so that mutation, crossover, and archive-based mutation are engaged at probabilities 0.2, 0.4, and 0.4, respectively.

archive-based mutation

Table 10.5 presents the results for the configurations with archives (called hereafter 'archived' configurations), marked with an 'A' appended to the list of search drivers. The success rates are remarkably higher than for configurations reported in Table 10.3. Overall, the configurations rank as follows:

FLEA	FLCEA	FLCA	FLSCEA	FLSCA	FLA	FLSEA	FLSA
2.74	2.91	3.91	3.91	4.62	5.06	5.32	7.53

As before, program size is clearly not beneficial. More importantly however, we may now assess the impact of archives on performance. The joint ranking

Table 10.6: Average ranks of configurations that include and exclude particular search drivers (for archived configurations).

	L	C	E	S
Configurations including the driver	9.74	8.66	9.82	12.57
Configurations excluding the driver	15.06	12.45	12.09	9.59

of the archived and non-archived methods (including conventional GP, i.e. F and F+) is as follows:

FLCEA	FLEA	FLCA	FLSCEA	FLSCA	FLA	FLSEA	FLCE	FLSA	FLE
2.94	3.15	4	4.76	5.29	5.56	6.44	10.1	11.1	11.8

FLC	F+	FE	FLSCE	F	FL	FLSE	FLSC	FSE	FLS	FS
12.4	12.7	12.9	13.9	14.8	15.1	15.6	15.9	16.6	17.8	18.3

The improvements are unquestionable. The archived configurations achieve top ranks, and only one of them (FLSA) ranks worse than any other non-archived configuration (FLCE). The average rank for configurations that use an archive is 5.4, and 14.5 for those that do not. The top-ranking configurations, FLCA and FLCEA, are practically guaranteed to solve the easiest problems (success rate 1.0), work well for the more difficult ones, and manage to occasionally solve even the hardest MAJ8 benchmark, a feat none of the non-archived configurations managed to achieve. Most of the archived configurations regularly solve also PAR6, another difficult benchmark not solved by the non-archived configurations; FLEA, the other top-ranking configuration, solves this problem in almost every third run (success rate 0.30).

In terms of statistical significance, Friedman test is conclusive again ($p \approx 10^{-43}$). Post-hoc analysis shows that FLCEA's rank is significantly better than FLSA's and than all the subsequent configurations. The seven top configurations (up to FLSEA) are significantly better than F, and the top four configurations (up to FLSCEA) are better than F+. Let us emphasize that this statistical analysis is very conservative, given the weak nature of the non-parametric Friedman test.

Concerning the contributions of individual search drivers, we juxtapose the average ranks of groups of configurations in Table 10.6. Analogously to the ranks presented in Table 10.4, the positive contributions of L, C and E are evident; this is particularly clear for L, which boosts performance by over five ranks on average. Program size S is again detrimental, albeit not as much as before (note that, given 21 configurations, the raw ranks achieved on individual benchmarks range here in the interval $[1, 21]$, compared to $[1, 13]$ in the previous section).

The relatively high number of search drivers used in parallel (up to five in case of FLSCE and FLSCEA) does not seem to be harmful. Many configurations that feature four or five search drivers rank high. The average rank of configurations with five drivers is 9.33, compared to 9.38 and 12.25 for four and three drivers respectively. Let us reiterate that this is far from obvious: multiobjective methods like NSGA-II are known to lose performance with the growing number of objectives, as that causes solutions less likely to dominate other solutions. The diversifying nature of multi-faceted behavioral evaluation apparently managed to counteract this weakness.

The self-regulatory mechanism of discarding half of the archive in every generation proved effective. At least for the considered benchmarks, the archives did not tend to blow up. The average size of an archive at the end of a run varied from 100 to 331 subprograms (with average at 159), depending on problem and configuration. Thanks to that, the computational overhead of archive management was moderate: the runtimes on contemporary PCs exceeded five minutes per GP run only for the hardest benchmarks.

10.5 Importance of subprogram selection

The archiving mechanism introduces an additional flow of subprograms from the current population, through the archive, to the subsequent population (Fig. 8.1). Several decisions are made on that way: first about the selection of subprograms to be submitted to the archive, then about the removal of the least useful subprograms from the archive, and finally about the insertion of subprograms into programs. Given the nontrivial nature of this process, it is justified to ask: do the choices of subprograms made by the PANGEA's classifier matter at all?

To verify this hypothesis, we prepare a control setup that maintains all the settings of archived configurations (Sect. 10.4), but chooses the subprograms to be archived at random. Technically, the REPTree inducer is applied to an execution record as explained the previous section, resulting in a subset of k selected attributes. However, those attributes are discarded and k attributes are drawn from the execution record at random (without replacement). The subsequent stages proceed as before, i.e. the k subprograms indicated by the randomly selected attributes are submitted to the archive and undergo further processing. Therefore, the classifier is engaged here only in order to obtain k, i.e. to avoid arbitrarily setting this number.

Below we present the ranking on success rate, including all the original archived configurations and the corresponding configurations randomized in the above way, where the names of the latter are prepended with letter 'r'.

FLCEA	FLEA	FLCA	FLSCEA	FLSCA	FLA	FLSEA	rFLCEA
3	3.06	4.21	4.56	5.56	6.03	6.56	8.35

rFLEA	rFLCA	rFLA	rFLSCEA	FLSA	rFLSEA	rFLSCA	rFLSA
9.26	10.6	10.7	11.7	12	12.8	12.9	14.7

The outcome is unambiguous: all randomized configurations achieve worse rank than their non-randomized counterparts. Choosing subprograms according to classifier indication is beneficial. Apparently, such subprograms are more useful building blocks than subprograms chosen at random (even though the latter are also prioritized by utilities (10.3)). The capability of detecting meaningful behavioral patterns should be thus deemed critical.

10.6 Contextual search drivers

In the last experiment, we augment the configurations considered earlier with the implicit fitness sharing search driver f_{IFS}, marked by letter I. Note that this is the only contextual search driver (Sect. 9.3) considered here. The configurations are simply the combinations of drivers used earlier, extended by I. We discard F, F+, FS, FE, and FSE, because they performed inferior in the earlier experiments.

For clarity, we report separately the ranks of non-archived combinations:

FLCE	FLEI	FLE	FLCEI	FLC	FLCI	FLSCE	FLI
3.82	5.15	5.29	5.85	6.09	7.94	8.35	8.94

FLSCEI	FL	FLSEI	FLSCI	FLSE	FLSC	FLSI	FLS
9.03	9.09	10.1	10.1	10.4	10.5	12.4	13

and the archived ones:

FLEA	FLCEA	FLCEAI	FLSCEA	FLCA	FLSCEAI	FLEAI	FLSCA
8.5	8.85	9.24	9.29	10.1	10.3	12.2	14.7

FLSCAI	FLCAI	FLA	FLSEAI	FLSEA	FLAI	FLSAI	FLSA
4.5	4.88	6.53	6.62	6.82	7.44	7.65	8.41

For the non-archived configurations, the average rank of the combinations extended with driver I is 9.09 and 7.92 for those not including I. For the archived configurations, the analogous numbers are 8.85 and 8.16, respectively. Therefore, we may conclude that the IFS search driver I is not beneficial as an extender of combinations of search drivers considered here. Though Friedman test is conclusive again ($p \approx 10^{-12}$), none of the extended configurations is better than its non-extended variant according to post-hoc analysis.

This is however not to say that driver I is never helpful: it does improve performance in some configurations, for instance advancing the rank of FLE from 5.29 to 5.15. Nevertheless, it may occasionally deteriorate performance quite gravely, as in the case of FLEAI vs. FLEA. Note however that in absolute terms all archive-based configurations perform very well (cf. Table 10.5) and the differences in success rates between them are often minor.

To explain the low usefulness of I, we hypothesize that this driver may be unable to provide sufficient additional information compared to F, the conventional evaluation function included in each configuration. Therefore, we conducted an additional experiment, not reported here in detail for brevity, where the driver I *replaced* driver F in the configurations considered earlier, leading to search driver combinations ILCEA, ILSCEA, ILEA, ILSCA, ILCA, ILSEA, ILAI, and ILSA. No I-including combination provided better performance than the corresponding combination with F. This suggests that, at least within the considered suite of benchmarks, drivers F and I are mutually redundant.

10.7 Discussion

The experiments reported in this chapter, which in total engaged over 30 000 evolutionary runs, corroborate the results presented in [101] and [96]. The multi-faceted evaluation by means of diversified search drivers systematically leads to performance improvements. Problems that are very hard to solve using conventional GP become tractable when behavioral search drivers provide a richer account of solution characteristics. Simultaneous use of search drivers of conceptually different 'pedigrees' (like ML-based L and information theory-based E) proves particularly beneficial and forms an important new result to elaborate on in future. Detrimental influence of the only non-behavioral search driver considered here, program size S, may be interpreted as another argument in favor of behavioral drivers. However, we obviously should not hasten to deem all non-behavioral drivers useless based on this single case.

We find it particularly encouraging that behavioral assessment proves effective in configurations that substantially vary from our previous work, including different program representation (tree-based GP here vs. PushGP in [101]) and substantial departure from the setup used in [96]: simpler archive management and utility definition, different population initialization, and larger populations, to mention the most important differences. This clearly suggests that the behavioral paradigm is robust and powerful enough to work well in various operating conditions.

In addition to the experiments reported here in detail, we tested also other variations, including:

- Incremental subprogram archives, i.e. merging the subprograms harvested from programs with the previous archive content,
- Drawing the subprograms from the archive with probabilities proportional to their utilities,

- Even simpler definition of feature utility, not discounted by the number of identified features, i.e. $u(p') = 1 - e(p)$ (vs. (10.3)).

In all cases, the comparisons of these variants with the core algorithm used in this chapter were inconclusive or revealed only minor differences, which inclined us to not present them in detail. On the other hand, this stability of performance is yet another signal that behavioral search drivers did not perform so well here by sheer luck, which corroborates the rationale behind behavioral program synthesis.

11

Implications of the behavioral perspective

Previous chapters presented a range of approaches that characterize program behavior in terms of execution record and search drivers. The experiments reported in Chap. 10 demonstrated that these approaches increase the likelihood of synthesizing a correct program. What are the other, not necessarily empirical, implications of behavioral program synthesis? We discuss them in a broader context in this chapter.

11.1 Conceptual consequences

When discussing advantages of behavioral evaluation in Sect. 9.8 and elsewhere, we suggested that it can be a means for assessing and promoting diversity in populations of programs. This is particularly natural in the presence of multiple search drivers. If neither of two compared programs dominates the other, selection renders them incomparable and allows them co-exist in a population. No additional mechanism for controlling or inducing diversity may be necessary. Behavioral evaluation and selection *implicitly* provide for phenotypic diversity, which in turn may lower the risk of premature convergence and overfocusing on local optima.

diversity maintenance

One might argue that maintaining behavioral diversity is not a truly novel feature, especially given the recent works addressing semantic diversity (e.g., [35]) or methods like lexicase selection [50]. However, the concept of an execution record invites a deeper take on diversity, not limited only to inspecting program output. For instance, two programs could be treated as behaviorally distinct if there is *any* difference in their execution records. Because program fragments are being constantly moved by crossover between individuals in the population, such a mechanism could promote 'internal behavioral diversity' and have positive impact on search performance. This supposition remains to be verified in further studies.

© Springer International Publishing Switzerland 2016
K. Krawiec, *Behavioral Program Synthesis with Genetic Programming*,
Studies in Computational Intelligence 618,
DOI: 10.1007/978-3-319-27565-9_11

Many of the search drivers considered in this book are in a sense *invented* by a search algorithm. By relying on ML-induced behavioral models, PANGEA can autonomously detect behavioral patterns that reveal potentially useful candidate solutions and parts thereof. No background knowledge or human ingenuity is necessary for that purpose: the experimenter is not required to specify which types of patterns are desirable. This makes PANGEA attractive when compared to, e.g., extensions of reinforcement learning methods (Sect. 9.10) that require the additional search drivers to be explicitly provided [8, 165].

Another, potentially more consequential feature of behavioral evaluation is facilitation of *problem decomposition*. Problem decomposition, often considered together with *modularity*, has been for long considered an important aspect of intelligent systems, and it remains to be an area of intense research in computational and artificial intelligence [190]. In behavioral program synthesis, there are at least two alternative avenues to problem decomposition.

By holding a trace for every test, execution records open the door to *vertical*, row-wise, test-wise problem decomposition. This capability is essential for geometric semantic GP (Chap. 5), where the geometric crossover operators combine the behaviors of parents on particular tests. For instance, in the exact geometric crossover in the Boolean domain (Fig. 5.3a and (5.8)), a mixing random subprogram decides, for each test individually, which parent to copy the output from. According to the convention adopted in this book, such outputs are column vectors (see, e.g., Figs. 7.5), and a mixing program, by picking elements from such vectors, effectively splice them into smaller vertical segments – hence the name.

Execution records facilitate also *horizontal*, column-wise problem decomposition, i.e. along the course of program execution. In the median example in Fig. 7.2, horizontal decomposition consists of splitting the original task into two separate subtasks of (i) sorting the list and (ii) retrieving the central element of the sorted list (or averaging the pair of the central elements for the even-length lists). We argue that such desirable decompositions can be automatically derived by analyzing execution records for entire *populations* of programs.

To proceed with our argument, we need to consider the joint behavioral space of multiple programs. Figure 11.1 visualizes the behavior of three programs that start with the same input and end up with the same output. Contrary to previous figures, where a graph node corresponded one-to-one to a state of an execution environment, here it represents a *combination* of execution states *for all tests* in T (a $|T|$-ary Cartesian product of execution states). We term such a combination a *c-state*. For instance, s_1 is the combination of executions states reached by program p_1 after executing its first instruction for all available tests. Consistently, a path in the graph represents a *combined trace* (*c-trace*) and captures the behavior of a program for all tests. c-traces

[margin notes:] problem decomposition modularity vertical problem decomposition horizontal problem decomposition

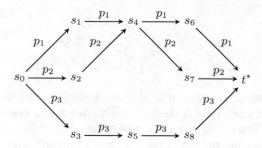

Fig. 11.1: Graphical representation of three c-traces of programs p_1, p_2, p_3, each of length 4. Each c-state s_i corresponds to a specific combination of states for all tests. Vertical columns of c-states correspond to consecutive execution stages.

of two programs can meet in a single c-state and then diverge (which was impossible for traces of the same program; see discussion in Sect. 6.2). In Fig. 11.1, this is the case for the c-state s_4 where the c-traces of p_1 and p_2 meet.

Assume for the sake of argument that p_1, p_2, and p_3 in Fig. 11.1 are correct programs that calculate the median. p_1 and p_2 may implement different sorting algorithms (which lead to some intermediate c-states s_1 and s_2), but they arrive at the same intermediate c-state s_4 where the list is sorted for all considered tests. The programs may then use different means to fetch the central element from the sorted list, which makes their c-traces diverge again. Nevertheless, they both end up in the same final c-state t^* (the target of the task), producing the correct value of the median. The fact that the c-trace of p_3 does not pass through s_4 could mean in this context that p_3 sorts the list in reverse order, or determines the median without sorting the list altogether.

The key observation related to horizontal decomposition is that the larger the number of correct programs with c-traces crossing in the same intermediate c-state s, the more evident it is that a task is decomposable. It is so because, on one hand, s can be achieved in many ways from the initial c-state, and, on the other, there are many ways of transforming s into the correct final c-state (target) t^*. We visualize such scenario for multiple programs in Fig. 11.2. If we knew s in advance, the task could be elegantly decomposed into two separate tasks. Crucially – and this is the main promise of problem decomposition – the combined computational effort of solving both these subtasks could be lower than that of solving the original task: the expected length of a correct program for each subtask is a fraction of the expected length of a program that solves the entire task, and the size of program space that needs to be searched depends exponentially on program length.

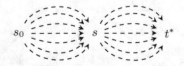

Fig. 11.2: The set of c-traces of a program synthesis task that is modular in the horizontal sense (the dashed lines denote c-traces traversing through many c-states). The c-traces traverse the same intermediate c-state s, while the computation that precede and follow that c-state can be realized in multiple ways (hence multiple arcs from s_0 to s and from s to t^*).

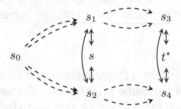

Fig. 11.3: The c-traces of close-to-correct programs cluster in the space of c-states (dashed arrows represent c-paths, solid lines similarity between c-states). The programs end execution in c-states s_3 and s_4 that are similar to target t^* (similarity marked with blue arrows). This may suggest that the intermediate c-states they traverse (s_1 and s_2, respectively) can be also similar in some sense, and indicate the existence of a (yet unknown) c-state s that could make it possible to achieve the target t^*.

In practice however such desirable c-states are not known in advance and, to make things harder, we do not even know if they exist for a given program synthesis task. Not discouraged by this, we posited in [87] that for horizontally modular tasks the c-traces of close-to-correct programs tend to visit *similar* intermediate c-states. Assume that the c-states s and t^* in Fig. 11.3 correspond to, respectively, the sorted list and the calculated median, as in the previous figures. Consider a program that fails to sort (and consequently produces wrong output) for some tests in T, with c-trace marked by any sequence of dashed arrows. Obviously, because the program passes quite many tests, its final c-state (say, s_3) will be similar to the target t^*. More importantly however, an intermediate c-state s_1 traversed by that program after (imperfect) sorting phase will share many execution states with the desired (yet in general unknown) intermediate c-state s, be in this sense similar[1] to it.

Given many programs in a working population, each of them possibly diverging from s in an individual way, the 'behavioral trajectories' of such

----

[1] The similarity measure could be just the number of execution states shared by the compared c-states.

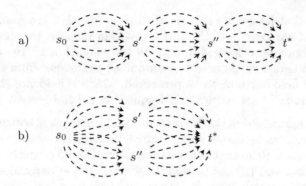

Fig. 11.4: The sets of c-traces for two program synthesis tasks, revealing three underlying subtasks (a), and indicating the possibility of realizing the target functionality via two alternative c-states (b).

close-to-correct programs may thus cluster, revealing so the internal behavioral structure of a task. In Fig. 11.3, such a cluster is formed by the c-traces traversing $s_0 \to s_1 \to s_3$ and $s_0 \to s_2 \to s_4$. Crucially, if both s_1 and s_2 are similar to s, then there is a chance that they are similar to each other, and an analogous observation holds for $s3$, s_4, and t^*. Therefore, clustering could be hypothetically observable even if s was not known. In [87], we provided preliminary experimental evidence for such clustering in real program synthesis tasks. We observed that the frequency with which c-traces cross with each other varies across tasks, and that the distribution of this characteristic across tasks is non-uniform, which corroborates the existence of less and more modular tasks in the above sense. In an earlier study [103], we showed that the proximity of the intermediate c-states correlates with the proximity to the target for some program synthesis tasks in the Boolean domain.

An effective method for detecting horizontal modularity would need to handle other ways in which c-traces could cluster. This includes more than two subtasks arranged sequentially (Fig. 11.4a) or subtasks arranged in parallel (Fig. 11.4b). These variants are clearly related to the concepts known in research on modularity [190]. A task for which all c-traces of correct programs intersect (Figs. 11.2 and 11.4a) is *separable*, because it can be decomposed into fully independent subtasks, each of them corresponding to a *module* and independent from all others. In evolutionary terms, there is no *epistasis* between such modules.

separable task

If only some c-traces of correct programs intersect, a task can be considered *nearly-decomposable* [163], or decomposable but not separable [190, p. 113]. For such tasks, changes within a part of program that corresponds to a module may or may not influence program behavior. In Fig. 11.4b, a subprogram with its c-trace ending in s' (a dashed arc from s_0 to s') may be

nearly-decomposable task

modified so that its c-trace still ends up in s' (another arc from s_0 to s') and the final outcome t^* of the entire program remains the same. But if the same subprogram changed so that its trace ended in s'', the remaining part of the program would need to be modified to traverse from s'' to t^*, in order to the final outcome to be preserved. There is evidence that nearly decomposable tasks are particularly frequent in natural systems [190].

The methods presented in this book do not perform explicit horizontal problem decomposition, i.e. do not appoint any intermediate c-states as desirable. However, those of them that define search drivers based on entire execution traces (PANGEA and TC) can be said to 'silently' aim at decomposition. For instance, the f_{TC} search driver (6.4) promotes programs (and thus indirectly also the corresponding c-states) characterized by high two-way entropy. In this way, it appoints as desirable an entire *class* of c-states rather than a *specific* c-state. An analogous comment applies to PANGEA (Chap. 7), where every attribute fetched by a classifier from an execution record (a column in Fig. 7.5) can be seen as a part of desired c-state. Telltales of problem decomposition become even more evident when combining these approaches with code reuse (Chap. 8): subprograms stored in an archive have c-traces that start in some s in the above figures and end in c-states that are desirable according to decisions made by a classifier in PANGEA.

semantic
back-
propa-
gation

There are alternatives to clustering when it comes to finding the desirable intermediate c-states. In [145], we proposed *semantic backpropagation*, where candidate programs are inversely executed (starting from t^*), producing by this means intermediate c-states that may be desirable. The key challenge of that approach is that inverse program execution is in general ambiguous, and many c-states may be thus appointed as potentially useful. Nevertheless, the empirical evidence presented in [145] clearly indicates viability of this approach and its superiority to alternative variants of semantic GP.

11.2 Architectural implications

The approaches of behavioral program synthesis presented in this book require extending the common EC framework with additional functionalities, like tracing program execution, multiobjective evaluation, and customized search operators (Chap. 7). It becomes thus natural to ask what are the implications of the behavioral paradigm for design of program synthesis methods and metaheuristics in general. In particular, does not it lead to tighter coupling between the components?

Let us start with noting that the two core concepts of our behavioral framework, execution record and search driver, are well-separated from the specifics of a given domain. An execution record is just a matrix[2] of values

---

[2] Or a two-level nested list, if traces vary in length.

that belong to the types available to the underlying programming language. It is essentially agnostic about program representation and programming language. For instance, an imperative program and a functional program may produce execution records that look very similar (or even identical, depending on the representations of execution states). Similarly, search drivers are largely problem-independent, and the behavioral ones require access to execution record only. Thus, although for clarity we defined them as functions of signature $\mathcal{P}^k \to \mathbb{O}^k$ (9.2), many of them could be redefined as $\mathcal{E}^k \to \mathbb{O}^k$, where $\mathcal{E}$ is the domain of execution records.

Nevertheless, before even bringing up the above arguments, we should start with stating that the presumed generic character of metaheuristic algorithms is largely a myth. The objective function is not the only component of the metaheuristic 'ecosystem' that is domain-specific and needs to be tailored to a given application. Consider the block diagram of a conventional, single-objective metaheuristic algorithm configured to solve a nontrivial program synthesis task (like GP in Fig. 1.1). Evaluation function features a nontrivial interpreter that runs programs on tests. There are sophisticated search operators that manipulate programs, possibly taking into account the specifics of a given programming language. Also the initialization phase needs to obey the admissible program syntax and/or type system. A selection mechanism may take into account not only program error, but also other properties like program size.

And between these all components we find the actual metaheuristic algorithm: usually just a handful of lines of code that control the flow of candidate solutions from domain-specific initialization to domain-specific evaluation to domain-dependent selection and domain-dependent search operators. Yet it is this algorithm which often, despite claiming to be generic, throttles the flow of information between the components mentioned above by, e.g., making narrowing assumptions about the nature of evaluation (e.g., assuming it to be scalar). The question is then: why should we insist on sticking to *meta*-heuristics and keeping them domain-agnostic, if the other, arguably more complex components, are often so much bound to problem domain? Opening the core search algorithm to the specifics of a given domain – effectively turning into a *heuristic* – may be easier than widely assumed, and very beneficial.

These observations urge us to diverge from the conventional evolutionary workflow and rethink the architecture of a behavioral program synthesis system. Rather than a rigid pipeline with predefined information flow, it may be more appropriate to conceptualize it as a *network of interconnected components* that exchange information about a search process. This vision fortuitously coincides with the recent analysis by Sörensen [168], where metaheuristics is first defined as an algorithmic *framework* accompanied with a set of guidelines (rather than an algorithm *template*), and later rephrased as follows:

> In its general sense, however, a metaheuristic is not an algorithm, i.e. it is not a sequence of actions that needs to be followed such as a cooking recipe. Rather, it is a consistent set of ideas, concepts, and operators that can be used to design heuristic optimization algorithms [168, p. 6].

The cited work also points out to advantages of this vision for research on metaheuristics: clearer understanding of past works, easier identification of similarities between the structure and inner workings of methods, and better focus on selected aspects of particular components. We hasten to extend that list by facilitation of hierarchical composition and higher abstraction levels.

separation of concerns

Component-based vision facilitates also *separation of concerns*, which in turn clarifies conceptualization and easies implementation. Decoupling evaluation from identifying the optimal solutions is an example of such separation in the proposed framework: a search driver provides only the former, while a correctness predicate only the latter. Separation of concerns is considered desirable in software engineering and sanctioned as a *design pattern* [36], i.e. a recognized good practice in software design. By analogy, the above separation of evaluation and correctness predicate can be seen as

metaheuristic design pattern

a *metaheuristic design pattern*[3]. Other components considered in this book, search drivers in particular, form viable candidates for such patterns too, which subscribes to the recent growing interest in metaheuristic standardization [174]. On a more technological level, this trend can be seen as a step toward seeing metaheuristic frameworks as service-oriented architectures[4].

Last but not least, the behavioral framework addresses the main problem identified in this book, i.e. that of *evaluation bottleneck* (Sect. 2.1). In the traditional GP workflow, an evaluation function plays the decisive role: it not only compresses the behavioral characteristics of a program into (usually scalar) evaluation (Sect. 2.1), but also decides *what* to compress. The 'client' components (selection operators and search operators) have no choice and need to make the best of what is available in such selective evaluation. With execution records, the role of evaluation is less dominant: it is responsible only for providing execution records, and leaves up to the client components how to exploit that information. And with more information available, the metaheuristic components become more empowered in their influence on search process. They can make better-informed decisions not only about selection of candidate solutions, but also about future search directions, e.g., by engaging search operators that reuse subprograms from an archive (Chap. 8). This decentralized 'whitebox' architecture interest-

blackboard architecture

ingly converges with the well-known *blackboard architecture* that has been known for decades in AI [156].

---

[3] See http://www.sigevo.org/gecco-2014/workshops.html#mdp
and http://www.sigevo.org/gecco-2015/workshops.html#wmdp.

[4] See http://osgiliath.org/ for an example of such an initiative.

11.3 Summary

In this chapter, we discussed only the most evident implications of adopting the behavioral perspective in program synthesis, focusing in particular on modularity, problem decomposition and architectural aspects. Concerning the latter, it is worth noting that some of architectural concerns can be elegantly addressed using the functional programming paradigm, which we demonstrate with our software suite [82] and elsewhere [174]. Many of the implications discussed in this chapter point to opportunities for future work, some of which we identify in the next, final chapter.

12

Future perspectives

This book proposes a new conceptual perspective on generate-and-test program synthesis. The framework of behavioral program synthesis is intended to provide more information on candidate programs on one hand, and to make search algorithms capable of exploiting that information on the other. The core elements of that framework are execution records, behavioral search drivers, multiobjective characterization, and code reuse. Experimental evidence in Chap. 10 and elsewhere proves this viable: behavioral insight into candidate programs is clearly beneficial. How can this take us further? What are the possible extensions? In this closing chapter, we outline a few prospective research directions.

12.1 The prospects

The repertoire of behavioral extensions of GP presented in this book is by no means complete. Our intention was to present the well-defined representatives of the behavioral paradigm and arrange them into a logical chain of gradually increasing sophistication. Such juxtaposition makes it easier to spot the opportunities for further work that can either interpolate or extrapolate this progression; in the following we point to several such possibilities.

The arguably most exciting aspect of search drivers is their vast design space. In this book, we only scratched the surface of possible designs. By its sheer size, the space of functions $\mathcal{P}^k \to \mathbb{O}^k$ must host many yet unknown search drivers. And even though only some of them can be expected to be universal, i.e. generic enough to prove effective for many problems (Sect. 9.5), and effective, i.e. guiding search better than the random search driver (Sect. 9.7), some may be more useful than anything known to date. Arguably, new behavioral patterns can be defined, detected and exploited,

© Springer International Publishing Switzerland 2016
K. Krawiec, *Behavioral Program Synthesis with Genetic Programming*,
Studies in Computational Intelligence 618,
DOI: 10.1007/978-3-319-27565-9_12

which would characterize other qualities than existence of execution traces that merge (TC, Chap. 6) or the capability of predicting the desired output (PANGEA, Chap. 7).

Interestingly, a search driver that has no obvious interpretation within our conceptual framework may still be very helpful. As the success of artificial neural networks shows, our ability to name or understand a concept (e.g., learned by a neuron in network's hidden layer) is not a prerequisite of performance. The repertoire of search drivers considered in this book can be surely extended with many other characteristics of program behavior. From the practical perspective, it would be particularly interesting to consider drivers reflecting non-functional aspects of program execution like program runtime, memory occupancy, or power consumption.

One of the features of search drivers that remains to be exploited is the *partial* nature of their codomains (9.2). All individual search drivers employed in the experimental part of this book order candidate programs completely, and partial ordering results only from their simultaneous use in a multiobjective setting. The key advantage of being partial is the permission to abstain from comparing some programs: for the success rate of a program synthesis algorithm, it may be better to decline such comparisons than to possibly suggest an incorrect ordering and so deceive an algorithm. In single-objective settings, partiality renders selection inconclusive and may be thus considered troublesome. This seems to be particularly reasonable in multiobjective configurations, where the other search drivers may fill in and relieve the momentarily 'incompetent' driver from its duties.

An aspect of program synthesis that is absent from this book is *data types*: we assumed single-type GP, as it is sufficient in quite many domains. Attempting to evolve programs that implement more complex concepts (i.e. operations on lists) necessitates multiple types. The importance of types has been advocated in program synthesis for a long time. Type constraints can immensely constrain the space of feasible solutions in program synthesis. For instance, as shown in [187], a function with the signature $List[T] \to \mathbb{N}$, where T is any type, must in fact be a function of list length, and only of list length (the reason being, among others, that the semantics of this task abstracts from the nature of list elements). There are no principal obstacles for designing search drivers that inspect types used by candidate programs and capture that information or confront it with the type-related information coming with the specific program synthesis task.

Richer behavioral information opens the door to a broader repertoire of possible reactions of a search algorithm. In this book, the recipients of behavioral evaluation are selection operators and, to a much lesser extent, search operators (the archive-based mutation in Chap. 8). Concerning other possibilities, generation of new candidate solutions can be more directional, as exemplified by semantic GP in Chap. 5, where search operators construct new candidate programs that meet very specific semantic requirements (see also [145]). The discussion on modularity in Sect. 11.1 points to possible extensions of the archive-based mutation. For instance, a behaviorally-aware crossover operator could involve *mate selection* and recombine the parents so as to achieve the desired behavioral effect. In an ideal scenario, such an operator, when faced with the task of synthesizing a program that calculates the median (Fig. 7.2) could select a sorting subprogram with a subprogram that retrieves the central element from a list, and combine them into a complete program. To an extent, we obtained this type effect in our recent work on *semantic backpropagation* [145], where the desired intermediate state is estimated via inverse execution of programs. Other components of GP workflow may benefit from behavioral evaluation as well; for instance, initializing programs with some form of feedback from evaluation process seems particularly promising.

To an extent, the idea of making search operators more responsive alludes to research on reactive search [9] and hyper-heuristics [18]. In both these research directions, a metaheuristic algorithm monitors its own progress and adjusts its search policy to maximize the odds of success. Their focus is on search operators and algorithms, i.e. on *how* to perform search. The behavioral perspective is more about *what* to drive the search with.

Concerning the impact beyond program synthesis, behavioral approach is applicable in domains where candidate solutions (agents) can interact with various 'stimuli' (environments) and produce diversified responses to them. The term *executable structure* has been quite often used in the past to characterize this class of solutions (Sect. 1.5.3). The behavioral perspective becomes even more fitting if the interactions between candidate solutions and environments are iterative: an agent reacts to a stimulus, takes an action which changes the environment state, and that change possibly affects the further course of its actions. A complete interaction involves multiple such stimulus-action cycles. Typical areas where such simulations are common include games, evolutionary robotics [15, 138], and, in general, AI agents situated in their environments.

The behavioral approach could push the envelope of these domains. As we argued in Sect. 1.1, by being capable of realizing any computable function, computer programs can in principle tackle all these tasks. Thus, in an interesting twist, one does not have to leave the realm of program synthesis to approach such problems. The past literature is abundant in examples of fruitful evolution of GP programs that control autonomous robots, solve complex problems, and play games (see [148] for a review).

In this book, we focused on behaviors of just one type of entity in a metaheuristic 'ecosystem' – programs. However, a metaheuristic algorithm itself exhibits certain behavior. In the limit, *any* aspects of search process can be the subject of behavioral analysis and exploitation. Like PANGEA peruses execution record, we might therefore hope to mine for patterns (behavioral or not) in, e.g., problem description, representation of solutions and operators, genotype-to-phenotype mapping, solution-state trajectory, algorithm-state trajectory, or operator-sequence trajectory. As an example, consider a set of tests $T \subseteq \mathbb{R}^2 \times \mathbb{R}$, i.e. for programs that take two real numbers as input and produce a real number in response, and assume that $((x, y), z) \in T \iff ((y, x), z) \in T$. Even without access to this domain knowledge, a human presented with sufficient samples from T would in all probability notice that the program to be induced should be symmetric with respect to its arguments. Taking this observation into account immensely reduces the size of search space. Such symmetry is of course a case of an *invariant*, and there are a number of ways in which one might attempt to incorporate invariants into heuristic program synthesis, perhaps the most obvious of which is to add it as a soft constraint of the heuristic function.

We expect the behavioral framework extended in the above ways to help advance heuristic program synthesis towards realizing the full potential discussed in Sect. 1.6 and help in scaling up already existing algorithms. The ML-style benchmarks traditionally studied in GP literature [122] may be insufficient to validate this claim, and the 'uncompromising' problems [50, 49] or the like might present a more appropriate challenge. Prospectively, we hope behavioral program synthesis to also be useful for conventional programming languages, with potential consequences for the practice of software engineering.

12.2 Closing remarks

We envision behavioral program synthesis as a paradigm shift from the traditional black-box architecture to the white-box one. In conventional GP, program behavior is hardly ever considered. Scalar evaluation, capturing the final effects of execution, is assumed to be informative enough to effectively drive search. We brought many arguments in this book, both theoretical and empirical, to claim that this is not the way to go. It should become clear at this point that evaluation in generative program synthesis can be made more informative at virtually no additional cost, and that the detailed information on program behavior is valuable and may substantially improve the performance of heuristic search.

In [168], Sörensen succinctly summarized Simon's and Newell's visionary paper on the role of heuristic methods [164], saying that

(...) a theory of heuristic (as opposed to algorithmic or exact) problem-solving should focus on intuition, insight, and learning. [168, p. 5]

The perspective presented in this book is parallel to that vision. Search drivers are in a sense *intuitive*, because they capture only certain aspects of considered candidate solutions, unlike objective functions that provide unquestionable assessment of solution quality (or are at least often believed to do so). A search driver may be even literally intuitive when it embodies a human's supposition about a presumably effective way of driving a search process. Concerning *insight*, search drivers clearly open the blackbox of evaluation function and make the structure of a problem more exposed to a metaheuristic search process. Finally, search drivers facilitate inclusion of *learning* in order to understand the structure of a search space (as in unsupervised learning via clustering in DOC) or to find the links between the behavioral characteristics of candidate solutions and the search goal (as in PANGEA).

One of the plausible interpretations of this book's motto "To measure is to know" is that science makes progress by developing better instruments. With this volume, we hope to extend the toolbox of instruments normally used in generate-and-test program synthesis. We look forward for further developments, including even more effective search drivers, a principled approach to designing search drivers, or, most excitingly, the dawn of automated design of search drivers. If hyper-heuristic research showed that design of metaheuristics can be automated, why should that be impossible for search drivers? Time will bring the answers to this and other suppositions formulated in and following from this volume.

Index

References

[1] Douglas Adams. *The Restaurant at the End of the Universe*. Pan Macmillan, 2009. ISBN: 9780330513111.

[2] Lee Altenberg. "Open Problems in the Spectral Analysis of Evolutionary Dynamics". In: *Frontiers of Evolutionary Computation*. Ed. by Anil Menon. Vol. 11. Genetic Algorithms And Evolutionary Computation Series. Boston, MA, USA: Kluwer Academic Publishers, 2004. Chap. 4, pp. 73–102. ISBN: 1-4020-7524-3. DOI: doi:10.1007/1-4020-7782-3_4. URL: http://dynamics.org/Altenberg/FILES/LeeOPSAED.pdf.

[3] Ignacio Arnaldo, Krzysztof Krawiec, and Una-May O'Reilly. "Multiple regression genetic programming". In: *GECCO '14: Proceedings of the 2014 conference on Genetic and evolutionary computation*. Ed. by Christian Igel et al. Vancouver, BC, Canada: ACM, Dec. 2014, pp. 879–886. DOI: doi:10.1145/2576768.2598291. URL: http://doi.acm.org/10.1145/2576768.2598291.

[4] Andrei Bajurnow and Vic Ciesielski. "Layered Learning for Evolving Goal Scoring Behavior in Soccer Players". In: *Proceedings of the 2004 IEEE Congress on Evolutionary Computation*. Portland, Oregon: IEEE Press, 20-23 06 2004, pp. 1828–1835. ISBN: 0-7803-8515-2. DOI: doi:10.1109/CEC.2004.1331118. URL: http://goanna.cs.rmit.edu.au/~vc/papers/cec2004-bajurnow.pdf.

[5] Wolfgang Banzhaf. "Genetic Programming and Emergence". In: *Genetic Programming and Evolvable Machines* 15.1 (Mar. 2014), pp. 63–73. ISSN: 1389-2576. DOI: doi:10.1007/s10710-013-9196-7.

[6] Wolfgang Banzhaf. "Genetic Programming for Pedestrians". In: *Proceedings of the 5th International Conference on Genetic Algorithms, ICGA-93*. Ed. by Stephanie Forrest. University of Illinois at Urbana-Champaign: Morgan Kaufmann, 17-21 07 1993, p. 628. URL: http://www.cs.ucl.ac.uk/staff/W.Langdon/ftp/ftp.io.com/papers/GenProg_forPed.ps.Z.

[7] Wolfgang Banzhaf et al. *Genetic Programming – An Introduction; On the Automatic Evolution of Computer Programs and its Applications*. San Francisco, CA, USA: Morgan Kaufmann, Jan. 1998. ISBN: 3-920993-58-6. URL: http://www.elsevier.com/wps/find/bookdescription.cws_home/677869/description#description.

[8] Andrew G. Barto, Satinder Singh, and Nuttapong Chentanez. "Intrinsically Motivated Learning of Hierarchical Collections of Skills". In: *Proceedings of International Conference on Developmental Learning (ICDL)*. Cambridge, MA: MIT Press, 2004.

[9] Roberto Battiti. "Reactive Search: Toward Self–Tuning Heuristics". In: *Modern Heuristic Search Methods*. Ed. by V. J. Rayward–Smith et al. Chichester: John Wiley & Sons Ltd., 1996, pp. 61–83.

[10] Lawrence Beadle and Colin Johnson. "Semantically Driven Crossover in Genetic Programming". In: *Proceedings of the IEEE*

World Congress on Computational Intelligence. Ed. by Jun Wang. IEEE Computational Intelligence Society. Hong Kong: IEEE Press, Jan. 2008, pp. 111–116. DOI: doi:10.1109/CEC.2008.4630784.

[11] Bir Bhanu, Yingqiang Lin, and Krzysztof Krawiec. *Evolutionary Synthesis of Pattern Recognition Systems.* Monographs in Computer Science. New York: Springer-Verlag, 2005. ISBN: 0-387-21295-7. URL: http://www.springer.com/west/home/computer/imaging?SGWID=4-149-22-39144807-detailsPage=ppmmedia%7CaboutThisBook.

[12] Christopher M. Bishop. *Pattern Recognition and Machine Learning (Information Science and Statistics).* Secaucus, NJ, USA: Springer-Verlag New York, Inc., 2006. ISBN: 0387310738.

[13] Yossi Borenstein and Alberto Moraglio. *Theory and Principled Methods for the Design of Metaheuristics.* Springer Publishing Company, Incorporated, 2014. ISBN: 3642332056, 9783642332050.

[14] Markus Brameier and Wolfgang Banzhaf. *Linear Genetic Programming.* Genetic and Evolutionary Computation XVI. Springer, 2007. ISBN: 0-387-31029-0. URL: http://www.springer.com/west/home/default?SGWID=4-40356-22-173660820-0.

[15] Rodney A. Brooks. *Cambrian Intelligence: The Early History of the New AI.* Cambridge, MA: MIT Press, Bradford Books, 1999 1999.

[16] Anthony Bucci, Jordan B. Pollack, and Edwin de Jong. "Automated Extraction of Problem Structure". In: *Genetic and Evolutionary Computation – GECCO-2004, Part I.* Ed. by Kalyanmoy Deb et al. Vol. 3102. Lecture Notes in Computer Science. Seattle, WA, USA: Springer-Verlag, 26-30 06 2004, pp. 501–512. ISBN: 3-540-22344-4. DOI: doi:10.1007/b98643. URL: http://link.springer.de/link/service/series/0558/bibs/3102/31020501.htm.

[17] Edmund K. Burke et al. "Towards the Decathlon Challenge of Search Heuristics". In: *Workshop on Automated Heuristic Design - In conjunction with the Genetic and Evolutionary Computation Conference (GECCO-2009), Montreal, Canada.* Montreal, Canada, 2009, pp. 2205–2208. URL: http://www.asap.cs.nott.ac.uk/publications/pdf/DecHH.pdf.

[18] Edmund Burke et al. "Hyper-Heuristics: An Emerging Direction in Modern Search Technology". English. In: *Handbook of Metaheuristics.* Ed. by Fred Glover and GaryA. Kochenberger. Vol. 57. International Series in Operations Research & Management Science. Springer US, 2003, pp. 457–474. ISBN: 978-1-4020-7263-5. DOI: 10.1007/0-306-48056-5_16. URL: http://dx.doi.org/10.1007/0-306-48056-5_16.

[19] Cristian Cadar, Daniel Dunbar, and Dawson Engler. "KLEE: Unassisted and Automatic Generation of High-coverage Tests for Complex Systems Programs". In: *Proceedings of the 8th USENIX Conference on Operating Systems Design and Implementation.* OSDI'08.

San Diego, California: USENIX Association, 2008, pp. 209–224. URL: http://dl.acm.org/citation.cfm?id=1855741.1855756.

[20] Mauro Castelli et al. "An Efficient Implementation of Geometric Semantic Genetic Programming for Anticoagulation Level Prediction in Pharmacogenetics". In: *Proceedings of the 16th Portuguese Conference on Artificial Intelligence, EPIA 2013*. Ed. by Luis Correia, Luis Paulo Reis, and Jose Cascalho. Vol. 8154. Lecture Notes in Computer Science. Angra do Heroismo, Azores, Portugal: Springer, Sept. 2013, pp. 78–89. DOI: doi:10.1007/978-3-642-40669-0_8. URL: http://link.springer.com/chapter/10.1007/978-3-642-40669-0_8.

[21] Siang Yew Chong et al. "Improving Generalization Performance in Co-Evolutionary Learning". In: *IEEE Transactions on Evolutionary Computation* 16.1 (2012), pp. 70–85. URL: http://www.cs.bham.ac.uk/~xin/papers/ChongTinoKuYaoTEVC2011.pdf.

[22] Peter Day and Asoke K. Nandi. "Binary String Fitness Characterization and Comparative Partner Selection in Genetic Programming". In: *IEEE Transactions on Evolutionary Computation* 12.6 (Dec. 2008), pp. 724–735. ISSN: 1089-778X. DOI: doi:10.1109/TEVC.2008.917201.

[23] Edwin D. de Jong and Anthony Bucci. "DECA: dimension extracting coevolutionary algorithm". In: *GECCO 2006: Proceedings of the 8th annual conference on Genetic and evolutionary computation*. Ed. by Mike Cattolico et al. Seattle, Washington, USA: ACM Press, 2006, pp. 313–320. ISBN: 1-59593-186-4. URL: http://doi.acm.org/10.1145/1143997.1144056.

[24] Edwin D. de Jong and Jordan B. Pollack. "Ideal Evaluation from Coevolution". In: *Evolutionary Computation* 12.2 (Summer 2004), pp. 159–192.

[25] Edwin D. de Jong, Richard A. Watson, and Jordan B. Pollack. "Reducing Bloat and Promoting Diversity using Multi-Objective Methods". In: *Proceedings of the Genetic and Evolutionary Computation Conference (GECCO-2001)*. Ed. by Lee Spector et al. San Francisco, California, USA: Morgan Kaufmann, July 2001, pp. 11–18. ISBN: 1-55860-774-9. URL: http://citeseer.ist.psu.edu/440305.html.

[26] Kalyanmoy Deb et al. "A fast and elitist multiobjective genetic algorithm: NSGA-II". In: *Evolutionary Computation, IEEE Transactions on* 6.2 (Apr. 2002), pp. 182–197. ISSN: 1089-778X. DOI: 10.1109/4235.996017.

[27] Edsger W. Dijkstra. "On the cruelty of really teaching computing science". circulated privately. Dec. 1988. URL: http://www.cs.utexas.edu/users/EWD/ewd10xx/EWD1036.PDF.

[28] Edsger W. Dijkstra. "On the reliability of programs". circulated privately. n.d. URL: http://www.cs.utexas.edu/users/EWD/ewd03xx/EWD303.PDF.

[29] M. Faifer, C. Janikow, and K. Krawiec. "Extracting fuzzy symbolic representation from artificial neural networks". In: *Proc. 18th In-*

ternational Conference of the North American Fuzzy Information Processing Society. New York, 1999, pp. 600–604.

[30]   Sevan G. Ficici. "Solution concepts in coevolutionary algorithms". Adviser-Pollack, Jordan B. PhD thesis. Waltham, MA, USA: Brandeis University, 2004.

[31]   Sevan G. Ficici and Jordan B. Pollack. "Pareto Optimality in Coevolutionary Learning". In: *Advances in Artificial Life, 6th European Conference, ECAL 2001*. Ed. by Jozef Kelemen and Petr Sosík. Vol. 2159. Lecture Notes in Computer Science. Prague, Czech Republic: Springer, 2001, pp. 316–325. ISBN: 3-540-42567-5. URL: http://link.springer.de/link/service/series/0558/bibs/2159/21590316.htm.

[32]   Jean-Christophe Filliâtre et al. *The Coq Proof Assistant - Reference Manual Version 6.1*. Tech. rep. 1997.

[33]   Lawrence Jerome Fogel, Alvin J. Owens, and Michael John Walsh. *Artificial Intelligence through Simulated Evolution*. New York: John Wiley, 1966.

[34]   Edgar Galvan-Lopez et al. "Defining locality as a problem difficulty measure in genetic programming". In: *Genetic Programming and Evolvable Machines* 12.4 (Dec. 2012), pp. 365–401. ISSN: 1389-2576. DOI: doi:10.1007/s10710-011-9136-3.

[35]   Edgar Galvan-Lopez et al. "Using Semantics in the Selection Mechanism in Genetic Programming: a Simple Method for Promoting Semantic Diversity". In: *2013 IEEE Conference on Evolutionary Computation*. Ed. by Luis Gerardo de la Fraga. Vol. 1. Cancun, Mexico, June 2013, pp. 2972–2979. DOI: doi:10.1109/CEC.2013.6557931.

[36]   Erich Gamma et al. *Design Patterns: Elements of Reusable Object-oriented Software*. Boston, MA, USA: Addison-Wesley Longman Publishing Co., Inc., 1995. ISBN: 0-201-63361-2.

[37]   Chris Gathercole and Peter Ross. "Dynamic Training Subset Selection for Supervised Learning in Genetic Programming". In: *Parallel Problem Solving from Nature III*. Ed. by Yuval Davidor, Hans-Paul Schwefel, and Reinhard Männer. Vol. 866. LNCS. Jerusalem: Springer-Verlag, Sept. 1994, pp. 312–321. ISBN: 3-540-58484-6. DOI: doi:10.1007/3-540-58484-6_275. URL: http://www.cs.ucl.ac.uk/staff/W.Langdon/ftp/papers/94-006.ps.gz.

[38]   Stuart Geman, Elie Bienenstock, and René Doursat. "Neural Networks and the Bias/Variance Dilemma". In: *Neural Comput.* 4.1 (Jan. 1992), pp. 1–58. ISSN: 0899-7667.

[39]   William F. Gilreath and Phillip A. Laplante. *Computer Architecture: A Minimalist Perspective: Dynamics and Sustainability*. The Springer International Series in Engineering and Computer Science. Springer US, 2003. ISBN: 9781402074165.

[40]   J. Gleick. *What Just Happened: A Chronicle from the Information Frontier*. Panthcon Books, 2002. ISBN: 9780375421778.

[41]   David Goldberg. *Genetic algorithms in search, optimization and machine learning*. Reading: Addison-Wesley, 1989.

[42] Ivo Goncalves and Sara Silva. "Balancing Learning and Overfitting in Genetic Programming with Interleaved Sampling of Training data". In: *Proceedings of the 16th European Conference on Genetic Programming, EuroGP 2013*. Ed. by Krzysztof Krawiec et al. Vol. 7831. LNCS. Vienna, Austria: Springer Verlag, Mar. 2013, pp. 73–84. DOI: doi:10.1007/978-3-642-37207-0_7.

[43] Ivo Goncalves et al. "Random Sampling Technique for Overfitting Control in Genetic Programming". In: *Proceedings of the 15th European Conference on Genetic Programming, EuroGP 2012*. Ed. by Alberto Moraglio et al. Vol. 7244. LNCS. Malaga, Spain: Springer Verlag, Nov. 2012, pp. 218–229. DOI: doi:10.1007/978-3-642-29139-5_19.

[44] Sumit Gulwani. "Dimensions in Program Synthesis". In: *Proceedings of the 12th international ACM SIGPLAN symposium on Principles and practice of declarative programming*. Invited talk. Hagenberg, Austria: ACM, Oct. 2010, pp. 13–24. DOI: doi:10.1145/1836089.1836091. URL: http://research.microsoft.com/en-us/um/people/sumitg/pubs/ppdp10-synthesis.pdf.

[45] Sumit Gulwani, William R. Harris, and Rishabh Singh. "Spreadsheet Data Manipulation Using Examples". In: *Communications of the ACM* 55.8 (Aug. 2012), pp. 97–105. ISSN: 0001-0782. DOI: doi:10.1145/2240236.2240260. URL: http://doi.acm.org/10.1145/2240236.2240260.

[46] Steven Gustafson and Leonardo Vanneschi. "Crossover-Based Tree Distance in Genetic Programming". In: *IEEE Transactions on Evolutionary Computation* 12.4 (Aug. 2008), pp. 506–524. ISSN: 1089-778X. DOI: doi:10.1109/TEVC.2008.915993.

[47] Mark Hall et al. "The WEKA Data Mining Software: An Update". In: *SIGKDD Explor. Newsl.* 11.1 (Nov. 2009), pp. 10–18. ISSN: 1931-0145. DOI: 10.1145/1656274.1656278. URL: http://doi.acm.org/10.1145/1656274.1656278.

[48] Thomas Haynes. "On-line Adaptation of Search via Knowledge Reuse". In: *Genetic Programming 1997: Proceedings of the Second Annual Conference*. Ed. by John R. Koza et al. Stanford University, CA, USA: Morgan Kaufmann, 13-16 07 1997, pp. 156–161. URL: http://citeseerx.ist.psu.edu/viewdoc/summary?doi=10.1.1.54.3381.

[49] Thomas Helmuth and Lee Spector. "General Program Synthesis Benchmark Suite". In: *Proceedings of the Genetic and Evolutionary Computation Conference, GECCO 2015, Madrid, Spain, July 11-15, 2015*. Ed. by Juan Luis Jiménez Laredo, Sara Silva, and Anna Isabel Esparcia-Alcázar. ACM, 2015, pp. 1039–1046. ISBN: 978-1-4503-3472-3. DOI: 10.1145/2739480.2754769. URL: http://doi.acm.org/10.1145/2739480.2754769.

[50] Thomas Helmuth, Lee Spector, and James Matheson. "Solving Uncompromising Problems with Lexicase Selection". In: *IEEE Trans-*

actions on Evolutionary Computation (). Accepted for future publication. ISSN: 1089-778X. DOI: doi:10.1109/TEVC.2014.2362729.

[51] Torsten Hildebrandt and Juergen Branke. "On Using Surrogates with Genetic Programming". In: *Evolutionary Computation* (). Forthcoming. ISSN: 1063-6560. DOI: doi:10.1162/EVCO_a_00133.

[52] Charles A. R. Hoare. "An Axiomatic Basis for Computer Programming". In: *Commun. ACM* 12.10 (Oct. 1969), pp. 576–580. ISSN: 0001-0782. DOI: 10.1145/363235.363259. URL: http://doi.acm.org/10.1145/363235.363259.

[53] Douglas R. Hofstadter. *Godel, Escher, Bach: An Eternal Golden Braid*. New York, NY, USA: Basic Books, Inc., 1979. ISBN: 0465026850.

[54] J.H. Holland. *Adaptation in natural and artificial systems*. Vol. 1. Ann Arbor: University of Michigan Press, 1975, pp. 75–89.

[55] John H. Holland. "Adaptation". In: *Progress in theoretical biology IV*. Ed. by R. Rosen and F. M. Snell. New York: Academic Press, 1976, pp. 263–293.

[56] John H. Holland. "Emergence". In: *Philosophica* 59.1 (1997), pp. 11–40.

[57] Myles Hollander and Douglas A. Wolfe. *Nonparametric Statistical Methods*. A Wiley-Interscience publication. Wiley, 1999. ISBN: 9780471190455.

[58] Gregory S. Hornby and Jordan B. Pollack. "Creating High-Level Components with a Generative Representation for Body-Brain Evolution". In: *Artif. Life* 8.3 (2002), pp. 223–246. ISSN: 1064-5462. DOI: doi:10.1162/106454602320991837. URL: http://www.demo.cs.brandeis.edu/papers/hornby_alife02.pdf.

[59] W. A. Howard. "The formulae-as-types notion of construction". In: *To H. B. Curry: essays on combinatory logic, lambda calculus and formalism*. Ed. by J. P. Seldin and J. R. Hindley. London-New York: Academic Press, 1980, pp. 480–490.

[60] Ting Hu et al. "Robustness, Evolvability, and Accessibility in Linear Genetic Programming". In: *Proceedings of the 14th European Conference on Genetic Programming, EuroGP 2011*. Ed. by Sara Silva et al. Vol. 6621. LNCS. Turin, Italy: Springer Verlag, 27-29 04 2011, pp. 13–24. DOI: doi:10.1007/978-3-642-20407-4_2.

[61] Hitoshi Iba, Taisuke Sato, and Hugo de Garis. "System identification approach to genetic programming". In: *Proceedings of the 1994 IEEE World Congress on Computational Intelligence*. Vol. 1. Orlando, Florida, USA: IEEE Press, 27-29 06 1994, pp. 401–406. DOI: doi:10.1109/ICEC.1994.349917.

[62] David Jackson. "Phenotypic Diversity in Initial Genetic Programming Populations". In: *Proceedings of the 13th European Conference on Genetic Programming, EuroGP 2010*. Ed. by Anna Isabel Esparcia-Alcazar et al. Vol. 6021. LNCS. Istanbul: Springer, July 2010, pp. 98–109. DOI: doi:10.1007/978-3-642-12148-7_9.

[63] Wojciech Jaśkowski. "Algorithms for Test-Based Problems". Adviser: Krzysztof Krawiec. PhD thesis. Poznan, Poland: Institute of Computing Science, Poznan University of Technology, 2011.

[64] Wojciech Jaśkowski and Krzysztof Krawiec. "Formal Analysis, Hardness and Algorithms for Extracting Internal Structure of Test-Based Problems". In: *Evolutionary Computation* 19.4 (2011), pp. 639–671. DOI: 10.1162/EVCO_a_00046. URL: http://www.mitpressjournals.org/doi/abs/10.1162/EVCO_a_00046.

[65] Wojciech Jaskowski, Krzysztof Krawiec, and Bartosz Wieloch. "Cross-Task Code Reuse in Genetic Programming Applied to Visual Learning". In: *International Journal of Applied Mathematics and Computer Science* 24.1 (2014), pp. 183–197. DOI: 10.2478/amcs-2014-0014. URL: http://www.cs.put.poznan.pl/kkrawiec/pubs/2013AMCS.pdf.

[66] Wojciech Jaskowski, Krzysztof Krawiec, and Bartosz Wieloch. "Genetic programming for cross-task knowledge sharing". In: *GECCO '07: Proceedings of the 9th annual conference on Genetic and evolutionary computation.* Ed. by Dirk Thierens et al. Vol. 2. London: ACM Press, July 2007, pp. 1620–1627. DOI: doi:10.1145/1276958.1277281. URL: http://www.cs.bham.ac.uk/~wbl/biblio/gecco2007/docs/p1620.pdf.

[67] Wojciech Jaskowski, Krzysztof Krawiec, and Bartosz Wieloch. "Multi-task code reuse in genetic programming". In: *GECCO-2008 Late-Breaking Papers.* Ed. by Marc Ebner et al. Atlanta, GA, USA: ACM, Dec. 2008, pp. 2159–2164. DOI: doi:10.1145/1388969.1389040. URL: http://www.cs.bham.ac.uk/~wbl/biblio/gecco2008/docs/p2159.pdf.

[68] Wojciech Jaskowski, Krzysztof Krawiec, and Bartosz Wieloch. "Multitask Visual Learning Using Genetic Programming". In: *Evolutionary Computation* 16.4 (Winter 2008), pp. 439–459. ISSN: 1063-6560. DOI: doi:10.1162/evco.2008.16.4.439.

[69] Wojciech Jaśkowski et al. "Improving coevolution by random sampling". In: *Proceeding of the fifteenth annual conference on Genetic and evolutionary computation conference.* GECCO '13. Amsterdam, The Netherlands: ACM, 2013, pp. 1141–1148. ISBN: 978-1-4503-1963-8. DOI: 10.1145/2463372.2463512. URL: http://doi.acm.org/10.1145/2463372.2463512.

[70] Mikkel T. Jensen. "Helper-objectives: Using multi-objective evolutionary algorithms for single-objective optimisation". In: *J. Math. Model. Algorithms* 3.4 (2004), pp. 323–347. DOI: 10.1007/s10852-005-2582-2. URL: http://dx.doi.org/10.1007/s10852-005-2582-2.

[71] Yaochu Jin, Markus Olhofer, and Bernhard Sendhoff. "A Framework for Evolutionary Optimization with Approximate Fitness Functions". In: *IEEE TRANSACTIONS ON EVOLUTIONARY COMPUTATION* 6 (2002), pp. 481–494.

[72] Gopal K. Kanji. *100 Statistical Tests*. SAGE Publications, 1999.
ISBN: 9780761961512.

[73] Karthik Kannappan et al. "Analyzing a Decade of Human-Competitive (HUMIE) Winners:What Can We Learn?" In: *Genetic Programming Theory and Practice XII*. Genetic and Evolutionary Computation. In preparation. Ann Arbor, USA: Springer, May 2014.

[74] Nadav Kashtan and Uri Alon. "Spontaneous evolution of modularity and network motifs". In: *Proceedings of the National Academy of Sciences* 102.39 (Sept. 2005), pp. 13773–13778. DOI: doi:10.1073/pnas.0503610102.   URL: http://www.pnas.org/cgi/reprint/102/39/13773.pdf.

[75] Nadav Kashtan, Elan Noor, and Uri Alon. "Varying environments can speed up evolution". In: *Proceedings of the National Academy of Sciences* 104.34 (21 08 2007), pp. 13711–13716. DOI: doi:10.1073/pnas.0611630104.   URL: http://www.pnas.org/cgi/reprint/104/34/13711.

[76] Joshua D. Knowles, Richard A. Watson, and David Corne. "Reducing Local Optima in Single-Objective Problems by Multi-objectivization". In: *EMO '01: Proceedings of the First International Conference on Evolutionary Multi-Criterion Optimization*. London, UK: Springer-Verlag, 2001, pp. 269–283. ISBN: 3-540-41745-1.

[77] Zoltan A. Kocsis and Jerry Swan. "Asymptotic Genetic Improvement Programming via Type Functors and Catamorphisms". In: *Semantic Methods in Genetic Programming*. Ed. by Colin Johnson et al. Workshop at Parallel Problem Solving from Nature 2014 conference. Ljubljana, Slovenia, 13 09 2014. URL: http://www.cs.put.poznan.pl/kkrawiec/smgp2014/uploads/Site/Kocsis.pdf.

[78] John R. Koza. *Genetic Programming II: Automatic Discovery of Reusable Programs*. Cambridge Massachusetts: MIT Press, May 1994. ISBN: 0-262-11189-6. URL: http://mitpress.mit.edu/catalog/item/default.asp?ttype=2&tid=8307.

[79] John R. Koza. *Genetic Programming: On the Programming of Computers by Means of Natural Selection*. Cambridge, MA, USA: MIT Press, 1992. ISBN: 0-262-11170-5. URL: http://mitpress.mit.edu/books/genetic-programming.

[80] John R. Koza. "Human-competitive results produced by genetic programming". In: *Genetic Programming and Evolvable Machines* 11.3/4 (Sept. 2010). Tenth Anniversary Issue: Progress in Genetic Programming and Evolvable Machines, pp. 251–284. ISSN: 1389-2576. DOI: doi:10.1007/s10710-010-9112-3. URL: http://www.genetic-programming.com/GPEM2010article.pdf.

[81] John R. Koza et al. *Genetic Programming IV: Routine Human-Competitive Machine Intelligence*. Kluwer Academic Publishers, 2003. ISBN: 1-4020-7446-8. URL: http://www.amazon.com/

Genetic-Programming-IV-Human-Competitive-Intelligence/dp/ 1402074468.

[82]   Krzysztof Krawiec. *Behavioral Program Synthesis with Genetic Programming (accompanying material)*. http://www.cs.put.poznan.pl/ kkrawiec/bps/. [Online]. 2015.

[83]   Krzysztof Krawiec. *Evolutionary Feature Programming: Cooperative learning for knowledge discovery and computer vision*. 385. Poznan University of Technology, Poznan, Poland: Wydawnictwo Politechniki Poznanskiej, 2004. URL: http://www.cs.put.poznan.pl/ kkrawiec/pubs/hab/krawiec_hab.pdf.

[84]   Krzysztof Krawiec. "Genetic programming: where meaning emerges from program code". In: *Genetic Programming and Evolvable Machines* 15.1 (Mar. 2014), pp. 75–77. ISSN: 1389-2576. DOI: doi:10.1007/s10710-013-9200-2.

[85]   Krzysztof Krawiec. "Learnable Embeddings of Program Spaces". In: *Proceedings of the 14th European Conference on Genetic Programming, EuroGP 2011*. Ed. by Sara Silva et al. Vol. 6621. LNCS. Turin, Italy: Springer Verlag, 27-29 04 2011, pp. 166–177. DOI: doi:10.1007/978-3-642-20407-4_15.

[86]   Krzysztof Krawiec. "Medial Crossovers for Genetic Programming". In: *Proceedings of the 15th European Conference on Genetic Programming, EuroGP 2012*. Ed. by Alberto Moraglio et al. Vol. 7244. LNCS. Malaga, Spain: Springer Verlag, Nov. 2012, pp. 61–72. DOI: doi:10.1007/978-3-642-29139-5_6.

[87]   Krzysztof Krawiec. "On relationships between semantic diversity, complexity and modularity of programming tasks". In: *GECCO '12: Proceedings of the fourteenth international conference on Genetic and evolutionary computation conference*. Ed. by Terry Soule et al. Philadelphia, Pennsylvania, USA: ACM, July 2012, pp. 783–790. DOI: doi:10.1145/2330163.2330272.

[88]   Krzysztof Krawiec. "Pairwise Comparison of Hypotheses in Evolutionary Learning". In: *Proceedings of the Eighteenth International Conference on Machine Learning (ICML 2001)*. Ed. by Carla E. Brodley and Andrea Pohoreckyj Danyluk. Williams College, Williamstown, MA, USA: Morgan Kaufmann, June 2001, pp. 266–273. ISBN: 1-55860-778-1. URL: http://citeseerx.ist.psu.edu/ viewdoc/summary?doi=10.1.1.29.900.pdf.

[89]   Krzysztof Krawiec and Bir Bhanu. "Visual Learning by Coevolutionary Feature Synthesis". In: *IEEE Transactions on System, Man, and Cybernetics – Part B* 35.3 (June 2005), pp. 409–425. DOI: doi:10.1109/TSMCB.2005.846644. URL: http://ieeexplore.ieee. org/iel5/3477/30862/01430827.pdf.

[90]   Krzysztof Krawiec and Bir Bhanu. "Visual Learning by Evolutionary and Coevolutionary Feature Synthesis". In: *IEEE Transactions on Evolutionary Computation* 11.5 (Oct. 2007), pp. 635–

650. DOI: doi:10.1109/TEVC.2006.887351. URL: http://ieeexplore. ieee.org/iel5/4235/4336114/04336120.pdf.

[91] Krzysztof Krawiec and Bir Bhanu. "Visual Learning by Evolutionary Feature Synthesis". In: *Proceedings of the Twentieth International Conference on Machine Learning (ICML 2003)*. Ed. by Tom Fawcett and Nina Mishra. Washington, DC, USA: AAAI Press, Aug. 2003, pp. 376–383. ISBN: 1-57735-189-4. URL: http://www.aaai.org/ Papers/ICML/2003/ICML03-051.pdf.

[92] Krzysztof Krawiec, Daniel Howard, and Mengjie Zhang. "Overview of Object Detection and Image Analysis by Means of Genetic Programming Techniques". In: *Proceedings of the 2007 International Conference Frontiers in the Convergence of Bioscience and Information Technologies (FBIT 2007)*. Jeju Island, Korea: IEEE Press, Oct. 2007, pp. 779–784. DOI: doi:10.1109/FBIT.2007.148.

[93] Krzysztof Krawiec and Pawel Lichocki. "Approximating geometric crossover in semantic space". In: *GECCO '09: Proceedings of the 11th Annual conference on Genetic and evolutionary computation*. Ed. by Guenther Raidl et al. Montreal: ACM, Aug. 2009, pp. 987–994. DOI: doi:10.1145/1569901.1570036.

[94] Krzysztof Krawiec and Pawel Lichocki. "Using Co-solvability to Model and Exploit Synergetic Effects in Evolution". In: *PPSN 2010 11th International Conference on Parallel Problem Solving From Nature*. Ed. by Robert Schaefer et al. Vol. 6239. Lecture Notes in Computer Science. Krakow, Poland: Springer, Nov. 2010, pp. 492–501. DOI: doi:10.1007/978-3-642-15871-1_50.

[95] Krzysztof Krawiec and Pawel Liskowski. "Automatic Derivation of Search Objectives for Test-Based Genetic Programming". In: *18th European Conference on Genetic Programming*. Ed. by Penousal Machado, Malcolm Heywood, and James McDermott. LNCS. Forthcoming. Copenhagen: Springer, Aug. 2015.

[96] Krzysztof Krawiec and Una-May O'Reilly. "Behavioral programming: a broader and more detailed take on semantic GP". In: *GECCO '14: Proceedings of the 2014 conference on Genetic and evolutionary computation*. Ed. by Christian Igel et al. Best paper. Vancouver, BC, Canada: ACM, Dec. 2014, pp. 935–942. DOI: doi:10.1145/2576768.2598288. URL: http://doi.acm.org/10.1145/ 2576768.2598288.

[97] Krzysztof Krawiec and Una-May O'Reilly. "Behavioral Search Drivers for Genetic Programing". In: *17th European Conference on Genetic Programming*. Ed. by Miguel Nicolau et al. Vol. 8599. LNCS. Granada, Spain: Springer, 23-25 04 2014, pp. 210–221. DOI: doi:10.1007/978-3-662-44303-3_18.

[98] Krzysztof Krawiec and Mikolaj Pawlak. "Genetic Programming with Alternative Search Drivers for Detection of Retinal Blood Vessels". In: *18th European Conference on the Applications of Evolutionary*

Computation. Ed. by Antonio Mora. LNCS. Forthcoming. Copenhagen: Springer, Aug. 2015.

[99]   Krzysztof Krawiec and Tomasz Pawlak. "Locally geometric semantic crossover: a study on the roles of semantics and homology in recombination operators". In: *Genetic Programming and Evolvable Machines* 14.1 (Mar. 2013), pp. 31–63. ISSN: 1389-2576. DOI: doi:10.1007/s10710-012-9172-7.

[100] Krzysztof Krawiec and Armando Solar-Lezama. "Improving Genetic Programming with Behavioral Consistency Measure". In: *13th International Conference on Parallel Problem Solving from Nature*. Ed. by Thomas Bartz-Beielstein et al. Vol. 8672. Lecture Notes in Computer Science. Ljubljana, Slovenia: Springer, 13-17 09 2014, pp. 434–443. DOI: doi:10.1007/978-3-319-10762-2_43.

[101] Krzysztof Krawiec and Jerry Swan. "Pattern-guided genetic programming". In: *GECCO '13: Proceeding of the fifteenth annual conference on Genetic and evolutionary computation conference*. Ed. by Christian Blum et al. Amsterdam, The Netherlands: ACM, June 2013, pp. 949–956. DOI: doi:10.1145/2463372.2463496.

[102] Krzysztof Krawiec and Bartosz Wieloch. "Automatic generation and exploitation of related problems in genetic programming". In: *IEEE Congress on Evolutionary Computation (CEC 2010)*. Barcelona, Spain: IEEE Press, 18-23 07 2010. DOI: doi:10.1109/CEC.2010.5586120.

[103] Krzysztof Krawiec and Bartosz Wieloch. "Functional modularity for genetic programming". In: *GECCO '09: Proceedings of the 11th Annual conference on Genetic and evolutionary computation*. Ed. by Guenther Raidl et al. Montreal: ACM, Aug. 2009, pp. 995–1002. DOI: doi:10.1145/1569901.1570037.

[104] W. B. Langdon. "How many Good Programs are there? How Long are they?" In: *Foundations of Genetic Algorithms VII*. Ed. by Kenneth A. De Jong, Riccardo Poli, and Jonathan E. Rowe. Published 2003. Torremolinos, Spain: Morgan Kaufmann, Apr. 2002, pp. 183–202. ISBN: 0-12-208155-2. URL: http://www.cs.ucl.ac.uk/staff/W.Langdon/ftp/papers/wbl_foga2002.ps.gz.

[105] W. B. Langdon and S. M. Gustafson. "Genetic Programming and Evolvable Machines: ten years of reviews". In: *Genetic Programming and Evolvable Machines* 11.3/4 (Sept. 2010). Tenth Anniversary Issue: Progress in Genetic Programming and Evolvable Machines, pp. 321–338. ISSN: 1389-2576. DOI: doi:10.1007/s10710-010-9111-4. URL: http://www.cs.ucl.ac.uk/staff/W.Langdon/ftp/papers/gppubs10.pdf.

[106] W. B. Langdon and Riccardo Poli. *Foundations of Genetic Programming*. Springer-Verlag, 2002. ISBN: 3-540-42451-2. DOI: doi:10.1007/978-3-662-04726-2.   URL: http://link.springer.com/book/10.1007/978-3-662-04726-2.

[107]   William B. Langdon and Mark Harman. "Genetically Improved CUDA C++ Software". In: *17th European Conference on Genetic Programming*. Ed. by Miguel Nicolau et al. Vol. 8599. LNCS. Granada, Spain: Springer, 23-25 04 2014, pp. 87–99. DOI: doi:10.1007/978-3-662-44303-3_8. URL: http://www.cs.ucl.ac.uk/staff/W.Langdon/ftp/papers/langdon_2014_EuroGP.pdf.

[108]   William B. Langdon et al. "Improving 3D Medical Image Registration CUDA Software with Genetic Programming". In: *GECCO '14: Proceeding of the sixteenth annual conference on genetic and evolutionary computation conference*. Ed. by Christian Igel et al. Vancouver, BC, Canada: ACM, Dec. 2014, pp. 951–958. DOI: doi:10.1145/2576768.2598244. URL: http://doi.acm.org/10.1145/2576768.2598244.

[109]   Christian W. G. Lasarczyk, Peter Dittrich, and Wolfgang Banzhaf. "Dynamic Subset Selection Based on a Fitness Case Topology". In: *Evolutionary Computation* 12.2 (Summer 2004), pp. 223–242. DOI: doi:10.1162/106365604773955157. URL: http://ls11-www.cs.uni-dortmund.de/people/lasar/publication/LasarDittBanz_TBS_2004/LasarDittBanz_TBS_2004.pdf.

[110]   Joel Lehman and Kenneth O. Stanley. "Abandoning Objectives: Evolution through the Search for Novelty Alone". In: *Evolutionary Computation* 19.2 (Summer 2011), pp. 189–223. ISSN: 1063-6560. DOI: doi:10.1162/EVCO_a_00025.

[111]   Joel Lehman and Kenneth O. Stanley. "Efficiently evolving programs through the search for novelty". In: *GECCO '10: Proceedings of the 12th annual conference on Genetic and evolutionary computation*. Ed. by Juergen Branke et al. Portland, Oregon, USA: ACM, July 2010, pp. 837–844. DOI: doi:10.1145/1830483.1830638. URL: http://eplex.cs.ucf.edu/papers/lehman_gecco10b.pdf.

[112]   Paweł Liskowski and Krzysztof Krawiec. "Discovery of Implicit Objectives by Compression of Interaction Matrix in Test-Based Problems". In: *Parallel Problem Solving from Nature – PPSN XIII*. Ed. by Thomas Bartz-Beielstein et al. Vol. 8672. Lecture Notes in Computer Science. Heidelberg: Springer, 2014, pp. 611–620. ISBN: 978-3-319-10761-5. DOI: 10.1007/978-3-319-10762-2_60.

[113]   Paweł Liskowski and Krzysztof Krawiec. "Online Discovery of Search Objectives for Test-based Problems". In: (). (under review).

[114]   Paweł Liskowski et al. "Comparison of Semantic-aware Selection Methods in Genetic Programming". In: *Proceedings of the Companion Publication of the 2015 on Genetic and Evolutionary Computation Conference*. GECCO Companion '15. Madrid, Spain: ACM, 2015, pp. 1301–1307. ISBN: 978-1-4503-3488-4. DOI: 10.1145/2739482.2768505. URL: http://doi.acm.org/10.1145/2739482.2768505.

[115]   Yi Liu and Taghi Khoshgoftaar. "Reducing overfitting in genetic programming models for software quality classification". In: *Proceedings*

of the Eighth IEEE Symposium on International High Assurance Systems Engineering. Tampa, Florida, USA, 25-26 03 2004, pp. 56–65. DOI: doi:10.1109/HASE.2004.1281730.

[116]   Darrell F Lochtefeld and Frank W Ciarallo. "Helper-objective optimization strategies for the job-shop scheduling problem". In: *Applied Soft Computing* 11.6 (2011), pp. 4161–4174.

[117]   Sean Luke. *Essentials of Metaheuristics.* First. Available at http://cs.gmu.edu/~sean/books/metaheuristics/.   lulu.com, 2009.   URL: http://www.lulu.com/shop/sean-luke/essentials-of-metaheuristics-second-edition/paperback/product-21080150.html.

[118]   Samir W. Mahfoud. "Niching methods for genetic algorithms". PhD thesis. Champaign, IL, USA, 1995.

[119]   John Maloney et al. "The Scratch Programming Language and Environment". In: *Trans. Comput. Educ.* 10.4 (Nov. 2010), 16:1–16:15. ISSN: 1946-6226. DOI: 10.1145/1868358.1868363. URL: http://doi.acm.org/10.1145/1868358.1868363.

[120]   Zohar Manna and Richard Waldinger. "A Deductive Approach to Program Synthesis". In: *ACM Trans. Program. Lang. Syst.* 2.1 (Jan. 1980), pp. 90–121. ISSN: 0164-0925. DOI: 10.1145/357084.357090. URL: http://doi.acm.org/10.1145/357084.357090.

[121]   Yuliana Martinez et al. "Searching for Novel Regression Functions". In: *2013 IEEE Conference on Evolutionary Computation.* Ed. by Luis Gerardo de la Fraga. Vol. 1. Cancun, Mexico, June 2013, pp. 16–23. DOI: doi:10.1109/CEC.2013.6557548. URL: http://eplex. cs.ucf.edu/noveltysearch/userspage/CEC-2013.pdf.

[122]   James McDermott et al. "Genetic programming needs better benchmarks". In: *GECCO '12: Proceedings of the fourteenth international conference on Genetic and evolutionary computation conference.* Ed. by Terry Soule et al. Philadelphia, Pennsylvania, USA: ACM, July 2012, pp. 791–798. DOI: doi:10.1145/2330163.2330273.

[123]   R. I. (Bob) McKay. "Committee Learning of Partial Functions in Fitness-Shared Genetic Programming". In: *Industrial Electronics Society, 2000. IECON 2000. 26th Annual Confjerence of the IEEE Third Asia-Pacific Conference on Simulated Evolution and Learning 2000.* Vol. 4. Nagoya, Japan: IEEE Press, Oct. 2000, pp. 2861–2866. ISBN: 0-7803-6456-2. DOI: doi:10.1109/IECON.2000.972452. URL: http://sc.snu.ac.kr/PAPERS/committee.pdf.

[124]   R I (Bob) McKay. "Fitness Sharing in Genetic Programming". In: *Proceedings of the Genetic and Evolutionary Computation Conference (GECCO-2000).* Ed. by Darrell Whitley et al. Las Vegas, Nevada, USA: Morgan Kaufmann, Oct. 2000, pp. 435–442. ISBN: 1-55860-708-0. URL: http://www.cs.bham.ac.uk/~wbl/biblio/ gecco2000/GP256.ps.

[125]   Nicholas Freitag McPhee, Brian Ohs, and Tyler Hutchison. "Semantic Building Blocks in Genetic Programming". In: *Proceedings of the 11th European Conference on Genetic Programming, EuroGP*

2008. Ed. by Michael O'Neill et al. Vol. 4971. Lecture Notes in Computer Science. Naples: Springer, 26-28 03 2008, pp. 134–145. DOI: doi:10.1007/978-3-540-78671-9_12.

[126]   P. Merz and B. Freisleben. "Fitness Landscape Analysis and Memetic Algorithms for the Quadratic Assignment Problem". In: *Trans. Evol. Comp* 4.4 (Nov. 2000), pp. 337–352. ISSN: 1089-778X. DOI: 10.1109/4235.887234. URL: http://dx.doi.org/10.1109/4235. 887234.

[127]   Bertrand Meyer. "Applying "Design by Contract"". In: *Computer* 25.10 (Oct. 1992), pp. 40–51. ISSN: 0018-9162. DOI: 10.1109/2.161279. URL: http://dx.doi.org/10.1109/2.161279.

[128]   George A. Miller. "Informavores". In: *The Study of Information: Interdisciplinary Messages*. Ed. by Una Machlup Fritz; Mansfield. Wiley-Interscience, 1983, pp. 111–113.

[129]   Julian F. Miller, ed. *Cartesian Genetic Programming*. Natural Computing Series. Springer, 2011. DOI: doi:10.1007/978-3-642-17310-3. URL: http://www.springer.com/computer/theoretical+computer+science/ book/978-3-642-17309-7.

[130]   Marvin Minsky. "Steps toward artificial intelligence". In: *Computers and Thought*. McGraw-Hill, 1961, pp. 406–450.

[131]   M. Mitchell. *Complexity: A Guided Tour*. OUP USA, 2009. ISBN: 9780195124415.

[132]   Melanie Mitchell. *An Introduction to Genetic Algorithms*. MIT Press, 1996. ISBN: 0-262-13316-4. URL: http://www.santafe.edu/ ~mm/books.html.

[133]   Richard Mitchell, Jim McKim, and Bertrand Meyer. *Design by Contract, by Example*. Redwood City, CA, USA: Addison Wesley Longman Publishing Co., Inc., 2002. ISBN: 0-201-63460-0.

[134]   T.M. Mitchell. *An introduction to genetic algorithms*. Cambridge, MA: MIT Press, 1996.

[135]   Alberto Moraglio, Krzysztof Krawiec, and Colin G. Johnson. "Geometric Semantic Genetic Programming". In: *Parallel Problem Solving from Nature, PPSN XII (part 1)*. Ed. by Carlos A. Coello Coello et al. Vol. 7491. Lecture Notes in Computer Science. Taormina, Italy: Springer, Sept. 2012, pp. 21–31. DOI: doi:10.1007/978-3-642-32937-1_3.

[136]   Quang Uy Nguyen. "Examining Semantic Diversity and Semantic Locality of Operators in Genetic Programming". PhD thesis. Ireland: University College Dublin, 18 07 2011. URL: http://ncra.ucd. ie/papers/Thesis_Uy_Corrected.pdf.

[137]   Jason Noble and Richard A. Watson. "Pareto coevolution: Using performance against coevolved opponents in a game as dimensions for Pareto selection". In: *Proceedings of the Genetic and Evolutionary Computation Conference (GECCO-2001)*. Ed. by Lee Spector et al. San Francisco, California, USA: Morgan Kaufmann, July 2001,

pp. 493–500. ISBN: 1-55860-774-9. URL: http://www.cs.bham.ac.uk/ ~wbl/biblio/gecco2001/d03b.pdf.

[138] Stefano Nolfi and Dario Floreano. *Evolutionary Robotics: The Biology,Intelligence,and Technology*. Cambridge, MA, USA: MIT Press, 2000. ISBN: 0262140705.

[139] Martin A. Nowak. *Evolutionary Dynamics: Exploring the Equations of Life*. Harvard University Press, 2006. URL: http://groups.lis.illinois.edu/amag/langev/paper/ nowak06evolutionaryDynamicsBOOK.html.

[140] Una-May O'Reilly. "Using a Distance Metric on Genetic Programs to Understand Genetic Operators". In: *IEEE International Conference on Systems, Man, and Cybernetics, Computational Cybernetics and Simulation*. Vol. 5. Orlando, Florida, USA, Dec. 1997, pp. 4092–4097. ISBN: 0-7803-4053-1. URL: http://ieeexplore.ieee.org/ iel4/4942/13793/00637337.pdf.

[141] Hirotugu Akaike. "A New Look at the Statistical Model Identification". In: *IEEE Transactions on Automatic Control* 19.6 (1974), pp. 716–723.

[142] Tomasz P. Pawlak. "Competent Algorithms for Geometric Semantic Genetic Programming". in review. PhD thesis. Pozna'n, Poland: Poznan University of Technology, 2015.

[143] Tomasz P. Pawlak and Krzysztof Krawiec. "Guarantees of Progress for Geometric Semantic Genetic Programming". Presented at Semantic Methods in Genetic Programming, PPSN'14 Workshop. 2014.

[144] Tomasz P. Pawlak, Bartosz Wieloch, and Krzysztof Krawiec. "Review and comparative analysis of geometric semantic crossovers". English. In: *Genetic Programming and Evolvable Machines* (). Online first. ISSN: 1389-2576. DOI: doi:10.1007/s10710-014-9239-8.

[145] Tomasz P. Pawlak, Bartosz Wieloch, and Krzysztof Krawiec. "Semantic Backpropagation for Designing Search Operators in Genetic Programming". In: *IEEE Transactions on Evolutionary Computation* (). This article has been accepted for publication in a future issue of this journal, but has not been fully edited. Content may change prior to final publication. ISSN: 1089-778X. DOI: doi:10.1109/TEVC.2014.2321259. URL: http://www.cs.put.poznan. pl/tpawlak/?Semantic%20Backpropagation%20for%20Designing %20Search%20Operators%20in%20Genetic%20Programming,16.

[146] Justyna Petke, William B. Langdon, and Mark Harman. "Applying Genetic Improvement to MiniSAT". In: *Symposium on Search-Based Software Engineering*. Ed. by Guenther Ruhe and Yuanyuan Zhang. Vol. 8084. Lecture Notes in Computer Science. Short Papers. Leningrad: Springer, Aug. 2013, pp. 257–262. DOI: doi:10.1007/978-3-642-39742-4_21. URL: ftp://ftp.cs.ucl.ac.uk/ genetic/papers/Petke_2013_SSBSE.pdf.

168   REFERENCES

[147]   Justyna Petke et al. "Using Genetic Improvement and Code Trans-
        plants to Specialise a C++ Program to a Problem Class". In:
        17th European Conference on Genetic Programming. Ed. by Miguel
        Nicolau et al. Vol. 8599. LNCS. Granada, Spain: Springer, 23-
        25 04 2014, pp. 137–149. DOI: doi:10.1007/978-3-662-44303-3_12.
        URL: http://www0.cs.ucl.ac.uk/staff/J.Petke/papers/Petke_2014_
        EuroGP.pdf.

[148]   Riccardo Poli, William B. Langdon, and Nicholas Freitag McPhee.
        A field guide to genetic programming. (With contributions by J.
        R. Koza). Published via http://lulu.com and freely available
        at http://www.gp-field-guide.org.uk, 2008. URL: http://dl.acm.
        org/citation.cfm?id=1796422.

[149]   Elena Popovici et al. "Handbook of Natural Computing". In: ed. by
        Grzegorz Rozenberg, Thomas Back, and Joost N. Kok. Springer-
        Verlag, 2011. Chap. Coevolutionary Principles.

[150]   John R. Quinlan. *C4.5: Programs for machine learning.* San Mateo:
        Morgan Kaufmann, 1992.

[151]   Ingo Rechenberg. *Evolutionsstrategie : Optimierung technischer Sys-
        teme nach Prinzipien der biologischen Evolution.* Problemata 15.
        Stuttgart-Bad Cannstatt: Frommann-Holzboog, 1973.

[152]   Jorma J. Rissanen. "Modeling By Shortest Data Description". In:
        Automatica 14 (1978), pp. 465–471.

[153]   Justinian P. Rosca and Dana H. Ballard. "Discovery of Subroutines
        in Genetic Programming". en. In: *Advances in Genetic Program-
        ming 2.* Ed. by Peter J. Angeline and K. E. Kinnear, Jr. Cam-
        bridge, MA, USA: MIT Press, 1996. Chap. 9, pp. 177–201. ISBN: 0-
        262-01158-1. URL: http://ieeexplore.ieee.org/xpl/articleDetails.jsp?
        tp=&arnumber=6277495.

[154]   Franz Rothlauf. *Representations for genetic and evolutionary algo-
        rithms.* Second. First published 2002, 2nd edition available electroni-
        cally. pub-SV:adr: Springer, 2006. ISBN: 3-540-25059-X. URL: http://
        download-ebook.org/index.php?target=desc&ebookid=5771.

[155]   Richard L. Rudell and Alberto L. Sangiovanni-Vincentelli.
        "ESPRESSO-MV: Algorithms for Multiple Valued Logic Min-
        imization". In: *Proc. of the IEEE Custom Integrated Circuits
        Conference.* Portland, May 1985.

[156]   Stuart J. Russell and Peter Norvig. *Artificial Intelligence: A Modern
        Approach.* 2nd ed. Pearson Education, 2003. ISBN: 0137903952.

[157]   Conor Ryan, Maarten Keijzer, and Mike Cattolico. "Favorable Bias-
        ing of Function Sets Using Run Transferable Libraries". In: *Genetic
        Programming Theory and Practice II.* Ed. by Una-May O'Reilly et
        al. Ann Arbor: Springer, 13-15 05 2004. Chap. 7, pp. 103–120. ISBN:
        0-387-23253-2. DOI: doi:10.1007/0-387-23254-0_7.

[158]   Eric Schkufza, Rahul Sharma, and Alex Aiken. "Stochastic
        Superoptimization". In: *Proceedings of the Eighteenth Interna-
        tional Conference on Architectural Support for Programming*

Languages and Operating Systems. ASPLOS '13. Houston, Texas, USA: ACM, 2013, pp. 305–316. ISBN: 978-1-4503-1870-9. DOI: 10.1145/2451116.2451150. URL: http://doi.acm.org/10.1145/2451116.2451150.

[159]   Ute Schmid and Fritz Wysotzki. "Skill Acquisition Can be Regarded as Program Synthesis". In: *In Mind Modelling - A Cognitive Science Approach to Reasoning, Learning and Discovery.* Pabst Science Publishers, 1998, pp. 39–45.

[160]   Hans-Paul Schwefel. "Kybernetische Evolution als Strategie der experimentellen Forschung in der Strömungstechnik". Diplomarbeit. Technische Universität Berlin, Hermann Föttinger–Institut für Strömungstechnik, März 1965.

[161]   Ehud Y. Shapiro. *Algorithmic Program DeBugging.* Cambridge, MA, USA: MIT Press, 1983. ISBN: 0262192187.

[162]   Ehud Y. Shapiro. *Inductive inference of theories from facts.* Tech. rep. RR 192. Yale University (New Haven, CT US), 1981. URL: http://opac.inria.fr/record=b1007525.

[163]   Herbert A. Simon. *The Sciences of the Artificial.* Cambridge, MA: MIT Press, 1969.

[164]   Herbert A. Simon and Allen Newell. "Heuristic problem solving: The next advance in operations research". In: *Operations research* 6.1 (1958), pp. 1–10.

[165]   Satinder P. Singh et al. "Intrinsically Motivated Reinforcement Learning: An Evolutionary Perspective". In: *IEEE Trans. on Auton. Ment. Dev.* 2.2 (June 2010), pp. 70–82. ISSN: 1943-0604. DOI: 10.1109/TAMD.2010.2051031. URL: http://dx.doi.org/10.1109/TAMD.2010.2051031.

[166]   Robert E. Smith, Stephanie Forrest, and Alan S. Perelson. "Searching for Diverse, Cooperative Populations with Genetic Algorithms". In: *Evolutionary Computation* 1.2 (June 1993), pp. 127–149. ISSN: 1063-6560. DOI: 10.1162/evco.1993.1.2.127. URL: http://dx.doi.org/10.1162/evco.1993.1.2.127.

[167]   Armando Solar-Lezama et al. "Combinatorial sketching for finite programs". In: *ASPLOS.* Ed. by John Paul Shen and Margaret Martonosi. ACM, 2006, pp. 404–415. ISBN: 1-59593-451-0.

[168]   Kenneth Sörensen. "Metaheuristics—the metaphor exposed". In: *International Transactions in Operational Research* 22.1 (2015), pp. 3–18. ISSN: 1475-3995. DOI: 10.1111/itor.12001. URL: http://dx.doi.org/10.1111/itor.12001.

[169]   Jonathan Sorg, Satinder P. Singh, and Richard L. Lewis. "Internal rewards mitigate agent boundedness". In: *Proceedings of the 27th international conference on machine learning (ICML-10).* 2010, pp. 1007–1014.

[170]   Lee Spector and Alan Robinson. "Genetic Programming and Autoconstructive Evolution with the Push Programming Language". In: *Genetic Programming and Evolvable Machines* 3.1 (Mar. 2002),

pp. 7–40. ISSN: 1389-2576. DOI: doi:10.1023/A:1014538503543. URL: http://hampshire.edu/lspector/pubs/push-gpem-final.pdf.

[171]   Lee Spector et al. "Genetic programming for finite algebras". In: *GECCO '08: Proceedings of the 10th annual conference on Genetic and evolutionary computation*. Ed. by Maarten Keijzer et al. Atlanta, GA, USA: ACM, Dec. 2008, pp. 1291–1298. DOI: doi:10.1145/1389095.1389343. URL: http://www.cs.bham.ac.uk/~wbl/biblio/gecco2008/docs/p1291.pdf.

[172]   Robert A. Stine. "Model Selection Using Information Theory and the MDL Principle". In: *Sociological Methods Research* 33.2 (Nov. 1, 2004), pp. 230–260. URL: http://smr.sagepub.com/cgi/content/abstract/33/2/230.

[173]   Richard S. Sutton and Andrew G. Barto. *Reinforcement Learning: An Introduction*. The MIT Press, 1998.

[174]   Jerry Swan et al. "A Research Agenda for Metaheuristic Standardization". In: *MIC 2015: The XI Metaheuristics International Conference*. (accepted). 2015.

[175]   Marcin Szubert. "Coevolutionary Shaping for Reinforcement Learning". PhD thesis. Poznan University of Technology, 2014. URL: http://www.cs.put.poznan.pl/mszubert/pub/phdthesis.pdf.

[176]   Marcin Szubert et al. "Shaping fitness function for evolutionary learning of game strategies". In: *Proceeding of the fifteenth annual conference on Genetic and evolutionary computation conference*. GECCO '13. Amsterdam, The Netherlands: ACM, 2013, pp. 1149–1156. ISBN: 978-1-4503-1963-8. DOI: 10.1145/2463372.2463513. URL: http://doi.acm.org/10.1145/2463372.2463513.

[177]   Walter A. Tackett and Aviram Carmi. "The unique implications of brood selection for genetic programming". In: *Proceedings of the 1994 IEEE World Congress on Computational Intelligence*. Vol. 1. Orlando, Florida, USA: IEEE Press, 27-29 06 1994, pp. 160–165. DOI: doi:10.1109/ICEC.1994.350023.

[178]   Xuejun Tan, Bir Bhanu, and Yingqiang Lin. "Fingerprint classification based on learned features". In: *IEEE Transactions on Systems, Man and Cybernetics, Part C: Applications and Reviews* 35.3 (Aug. 2005), pp. 287–300. ISSN: 1094-6977. DOI: doi:10.1109/TSMCC.2005.848167.

[179]   Fabien Teytaud and Olivier Teytaud. "Convergence rates of evolutionary algorithms and parallel evolutionary algorithms". In: *Theory and principled methods for the design of metaheuristics*. Ed. by Yossi Borenstein and Alberto Moraglio. Natural Computing Series. Springer, 2014, pp. 25–39. ISBN: 978-3-642-33205-0. DOI: 10.1007/978-3-642-33206-7_2.

[180]   *The On-Line Encyclopedia of Integer Sequences, published electronically at http://oeis.org*. 2011.

[181]   Marco Tomassini et al. "A Study of Fitness Distance Correlation as a Difficulty Measure in Genetic Programming". In:

Evolutionary Computation 13.2 (Summer 2005), pp. 213–239. ISSN: 1063-6560. DOI: doi:10.1162/1063656054088549. URL: http://www.ingentaconnect.com/search/expand;jsessionid=k43b8htgbpy4.victoria?pub=infobike://mitpress/evco/2005/00000013/00000002/art00004&unc=.

[182]   Alan M. Turing. "On Computable Numbers, with an Application to the Entscheidungsproblem". In: *Proceedings of the London Mathematical Society* s2-42.1 (1937), pp. 230–265. DOI: 10.1112/plms/s2-42.1.230. eprint: http://plms.oxfordjournals.org/content/s2-42/1/230.full.pdf+html. URL: http://plms.oxfordjournals.org/content/s2-42/1/230.short.

[183]   Nguyen Quang Uy et al. "On the roles of semantic locality of crossover in genetic programming". In: *Information Sciences* 235 (20 06 2013), pp. 195–213. ISSN: 0020-0255. DOI: doi:10.1016/j.ins.2013.02.008. URL: http://www.sciencedirect.com/science/article/pii/S0020025513001175.

[184]   Nguyen Quang Uy et al. "Semantically-based crossover in genetic programming: application to real-valued symbolic regression". In: *Genetic Programming and Evolvable Machines* 12.2 (June 2011), pp. 91–119. ISSN: 1389-2576. DOI: doi:10.1007/s10710-010-9121-2.

[185]   Pascal Vincent et al. "Stacked Denoising Autoencoders: Learning Useful Representations in a Deep Network with a Local Denoising Criterion". In: *J. Mach. Learn. Res.* 11 (Dec. 2010), pp. 3371–3408. ISSN: 1532-4435. URL: http://dl.acm.org/citation.cfm?id=1756006.1953039.

[186]   Michael D. Vose and Gunar Liepins. "Punctuated equilibria in genetic search". In: *Complex Systems* 5 (1991), pp. 31–44.

[187]   Philip Wadler. "Theorems for Free!" In: *Proceedings of the Fourth International Conference on Functional Programming Languages and Computer Architecture*. FPCA '89. Imperial College, London, United Kingdom: ACM, 1989, pp. 347–359. ISBN: 0-89791-328-0. DOI: 10.1145/99370.99404. URL: http://doi.acm.org/10.1145/99370.99404.

[188]   Richard J. Waldinger and Richard C. T. Lee. "PROW: A Step Toward Automatic Program Writing". In: *Proceedings of the 1st International Joint Conference on Artificial Intelligence, IJCAI*. Ed. by D. E. Walker and L. M. Norton. Morgan Kaufmann, 1969, pp. 241–252.

[189]   Henry S. Warren. *Hacker's Delight*. Boston, MA, USA: Addison-Wesley Longman Publishing Co., Inc., 2002. ISBN: 0201914654.

[190]   Richard A. Watson. *Compositional Evolution: The impact of Sex, Symbiosis and Modularity on the Gradualist Framework of Evolution*. Vol. NA. Vienna series in theoretical biology. MIT Press, Feb. 2006. URL: http://eprints.ecs.soton.ac.uk/10415/.

[191]   Darrell Whitley, Soraya Rana, and Robert B. Heckendorn. "The Island Model Genetic Algorithm: On Separability, Population Size

and Convergence". In: *Journal of Computing and Information Technology* 7.1 (1999), pp. 33–47.

[192]   David H. Wolpert and William G. Macready. "No Free Lunch Theorems for Optimization". In: *IEEE Trans. on Evolutionary Computation* 1.1 (1997), pp. 67–82.

[193]   David H. Wolpert and William G. Macready. *No Free Lunch Theorems for Search.* Tech. rep. SFI-TR-95-02-010. Santa Fe, NM: Santa Fe Institute, 1995.

[194]   Sewall Wright. "The roles of mutation, inbreeding, crossbreeding and selection in evolution". In: *Proc of the 6th International Congress of Genetics.* Vol. 1. 1932, pp. 356 366.

[195]   Wei Yan and Christopher D. Clack. "Behavioural GP diversity for adaptive stock selection". In: *GECCO '09: Proceedings of the 11th Annual conference on Genetic and evolutionary computation.* Ed. by Guenther Raidl et al. Montreal: ACM, Aug. 2009, pp. 1641–1648. DOI: doi:10.1145/1569901.1570120.

[196]   Wei Yan and Christopher D. Clack. "Behavioural GP diversity for dynamic environments: an application in hedge fund investment". In: *GECCO 2006: Proceedings of the 8th annual conference on Genetic and evolutionary computation.* Ed. by Maarten Keijzer et al. Vol. 2. Seattle, Washington, USA: ACM Press, Aug. 2006, pp. 1817–1824. ISBN: 1-59593-186-4. DOI: doi:10.1145/1143997.1144290. URL: http://www.cs.bham.ac.uk/~wbl/biblio/gecco2006/docs/p1817.pdf.

[197]   Byoung-Tak Zhang and Heinz Mühlenbein. "Balancing Accuracy and Parsimony in Genetic Programming". In: *Evolutionary Computation* 3.1 (1995), pp. 17–38. DOI: doi:10.1162/evco.1995.3.1.17. URL: http://www.ais.fraunhofer.de/~muehlen/publications/gmd_as_ga-94_09.ps.

Printed in the United States
By Bookmasters